Rewilding and Reintroductions in Britain

This book examines how rewilding and species reintroductions play a crucial role in conservation and ecological restoration in Britain.

The book also examines how humans think about and interact with our environment and nature, and how competing interests surrounding conservation, land use (in particular agriculture) and social-cultural issues can be reconciled. Rather than taking a stance which is aligned to either conservation or farming, as many other books do, this book examines how these two land uses are being negotiated in the discourse and reconciled in practice. Drawing on research involving visits to rewilding and reintroduction projects and interviews with conservation experts, practitioners and stakeholders, the book examines two landscapes sites (the Avalon Marshes in Somerset and Wild Ennerdale in Cumbria) and three species: cows, the European Wildcat and the Red Kite. It also examines the role played by public figures, whether activists, environmentalists or celebrities, in promoting rewilding. The book argues that rewilding in Britain is domesticated, in that it is more compatible with people and other land use than is often the case with rewilding in other parts of the world. This extends from the smaller scale at which it occurs, to the different names by which it is known, to the way that humans are still intervening in rewilding landscapes, and the very careful consideration which is being given to which animals are involved and how they are managed. Overall, this book provides insights into how rewilding in Britain is evolving with the potential impacts on land use and land use decision-making.

This book will be of great interest to students and scholars of rewilding, conservation, ecological restoration, human–wildlife interactions and environmental management more broadly.

Virginia Thomas is an environmental social scientist. She researches rewilding, human interactions with the environment, and human–animal relations. She holds a PhD in Sociology.

Routledge Studies in Conservation and the Environment

This series includes a wide range of inter-disciplinary approaches to conservation and the environment, integrating perspectives from both social and natural sciences. Topics include, but are not limited to, development, environmental policy and politics, ecosystem change, natural resources (including land, water, oceans and forests), security, wildlife, protected areas, tourism, human-wildlife conflict, agriculture, economics, law and climate change.

The Maya Forest Waterlands
Shared Conservation, Entangled Politics, and Fluid Borders
Hanna Laako and Edith Kauffer

Saving Biodiversity
Threats, Strategies, and Big Ideas
Matt W. Hayward

Rewilding and Ecological Justice
The Ethics and Politics of Wildlife Regeneration
Cristian Moyano-Fernández

Becoming Nature Positive
Transitioning to a Safe and Just Future
Marco Lambertini, Joseph W. Bull, Leroy Little Bear, Harvey Locke, Eva Zabey, Dorothy Maseke, and Carlos Manuel Rodríguez.

The Economics of the Wildlife Trade
Theory and Implications for Conservation
Frederick Chen and Michael't Sas-Rolfes

Rewilding and Reintroductions in Britain
Human Interactions with Conservation
Virginia Thomas

For more information about this series, please visit: www.routledge.com/Routledge-Studies-in-Conservation-and-the-Environment/book-series/RSICE

Rewilding and Reintroductions in Britain

Human Interactions with Conservation

Virginia Thomas

Routledge
Taylor & Francis Group
LONDON AND NEW YORK

First published 2026
by Routledge
4 Park Square, Milton Park, Abingdon, Oxon OX14 4RN

and by Routledge
605 Third Avenue, New York, NY 10158

*Routledge is an imprint of the Taylor & Francis Group,
an informa business*

Funded by Exeter University.

For Product Safety Concerns and Information please contact our EU representative GPSR@taylorandfrancis.com. Taylor & Francis Verlag GmbH, Kaufingerstraße 24, 80331 München, Germany.

British Library Cataloguing-in-Publication Data
A catalogue record for this book is available from the British Library

ISBN: 978-1-032-59834-5 (hbk)
ISBN: 978-1-032-59836-9 (pbk)
ISBN: 978-1-003-45649-0 (ebk)

DOI: 10.4324/9781003456490

Typeset in Optima
by SPi Technologies India Pvt Ltd (Straive)

This work was supported by the Wellcome Trust [219889_Z].

For my parents, for all the help you have given me in writing this book, directly and indirectly.
And with thanks to the friends who gave me words for this book, you know who you are.

Contents

Introduction

Rewilding and reintroductions in Britain: Human interactions with conservation

This book explores rewilding and reintroductions in Britain, the human and other-than-human animals involved, and the relationships between them. Some books have opening lines which require no explanation – this is not one of them. Introducing what this book explores requires the introduction and explanation of:

- Rewilding as a unique form of ecological restoration.
- Britain as a unique socio-ecological system with a unique application of rewilding.
- Species reintroductions and the other-than-human animals involved in rewilding.
- Human–animal relations and the biopolitics of rewilding.

These topics are introduced here, together with brief overviews of the chapters of this book, which explore the topics in more detail.

Two other points are important to note. First, that rewilding is still evolving rapidly as a conservation approach and trying to discuss it in relation to an entire country at book length means that the book will almost certainly be historical rather than current before it is finished. Rather than go back and start again, only to face the same challenge, this book accepts those limitations and should be viewed and used as a reflection of rewilding in Britain at the time it was written. Secondly, rewilding is a topic which prompts vociferous debate in academia, among its stakeholders, and in wider publics and media. This book is sure to provoke debate, not least because of the form of rewilding it describes as being present in Britain which is very different from rewilding in other parts of the world, particularly the United States. This book welcomes that debate and attempts to reflect rewilding in Britain as accurately as possible so as to support informed debate. It does not attempt to advocate for any of the approaches discussed, those are left to rewilding's practitioners and charismatic megafolk.

DOI: 10.4324/9781003456490-1

Rewilding as a unique form of ecological restoration

In 2021, the International Union for Conservation of Nature (IUCN) adopted a definition of rewilding as 'the process of rebuilding, following major human disturbance, a natural ecosystem by restoring natural processes and the complete or near complete food-web at all trophic levels as a self-sustaining and resilient ecosystem using biota that would have been present had the disturbance not occurred'. This definition did not exist when the research from which this book germinated began, which gives a hint of the rapidly evolving nature of the discourse around rewilding. In the absence of a universally agreed definition, and in the face of a plethora of competing, sometimes even conflicting, definitions, the research made no attempt to offer yet another definition. Instead, the concept of *family resemblance* (Wittgenstein, 1968) was used as a way of understanding rewilding and the different and diverse conservation approaches which adopt the term, or have it thrust upon them. Family resemblance is the idea that things may be grouped according to their possession of traits from a core set. Members of a group will be different yet similar, making them recognisable as members of the same group, because they each possess different combinations of these core traits, sharing some in common. Family resemblance is a particularly useful way to approach rewilding given that rewilding is a plastic term (Jørgensen, 2015), an umbrella term (Martin et al., 2021), and can mean different things to different people, in different places. Family resemblance was also a particularly useful way of exploring rewilding in the British context where it takes on a unique character.

Family resemblance is discussed in Chapter 1, as are the core set of traits which were identified as conferring a conservation approach with family resemblance to rewilding. These traits are identifying as rewilding, functioning at large scale, increasing biodiversity, increasing ecological functioning, reducing human intervention and increasing other-than-human agency. It is the latter two traits which distinguish rewilding from other types of ecological restoration which can be, and often is, highly interventionist. By contrast, rewilding seeks to reduce human intervention and, by the same token, to increase other-than-human agency. Rewilding is thus aiming for a fundamental shift in power from human to other-than-human, and to restore the control of landscapes and ecosystems to other-than-human processes.

Wildness, or other-than-human agency, is thus a fulcrum around which this book revolves. Wildness can be defined as 'the autonomy of the more-than-human world where events, such as animals moving about, plants growing, and rocks falling occur largely because of their own internal self-expression that is independent of civilized forces' (Woods, 2005, p. 177). In this definition, and in this book, wildness and agency (Woods uses the term autonomy) are interlinked, with an increase in agency corresponding with an increase in wildness. This is one of the major challenges for rewilding given that humans' lack of tolerance for other-than-human animals is commonly a lack of tolerance for their expression of agency (i.e., their wildness) (Murray, 2017). Given that rewilding calls for, and seeks, a reduction in human interventions in ecosystems (IUCN, 2021), it will necessitate a 'radical tolerance' from humans, not only of other species but of their wild agency (Campbell, 2006). Such tolerance is particularly problematic in Britain where there is a long history of human persecution and extirpation of other-than-human animals.

Britain as a unique socio-ecological system with a unique application of rewilding

Culturally and politically, Britain is three separate countries (England, Scotland and Wales), each of which has different approaches and attitudes to conservation, environmental policy and landscapes. Notwithstanding these differences, certain things can, and should, be noted about Britain as a whole since they are useful for our understanding of its cultural context. First, it has a long history of human habitation, and its landscapes are intensely cultural. Second, Britain considers itself as a nation of 'animal lovers' and a nation which cares about the environment. Third, hunting, and attitudes to it, are quite different from those in other parts of the world. These points are extremely significant in discourse surrounding conservation, particularly in relation to heritage landscapes, species conservation and animal welfare.

Geographically, Britain is an island, separated from mainland Europe circa 9000 years ago once the land bridge was submerged. It was connected to Europe again during the last ice age, allowing terrestrial animals to move back and forth across an ice bridge. The animals present on the island when the ice bridge melted at the start of the current interglacial period are those which are classified as native to Britain. Any animal which had been present previously, and yet was *not* present when the ice retreated, is not considered native. Animals which, for whatever reason, have disappeared from Britain *since* the ice retreated are considered extinct in Britain. Clearly, terrestrial mammals are now unable to recolonise Britain and any human intervention to restore them is classed as a reintroduction (reintroduction being defined as 'the deliberate or accidental release of a living organism(s) into the wild to areas where the species or race was native but has become extinct' (JNCC, 1996)). This is highly relevant when considering species conservation and reintroductions, for example, of the red kite (*Milvus milvus*) and the wildcat (*Felis sylvestris*), discussed in Chapters 6 and 7, respectively.

This book looks in detail at the rewilding landscapes of the Avalon Marshes and Wild Ennerdale, and more generally at the landscapes of reintroductions. Although very different, the landscapes of the Avalon Marshes and Wild Ennerdale share a long history of human habitation and modification. Similarly, sites of reintroduction across Britain, although on some levels 'wild' are still peopled landscapes. Visiting and walking in these landscapes in the years over which this research was conducted (2018–2023), talking with research participants, and taking notes and photographs built up a rich picture of the socioecological systems which are present and emerging in Britain as a result of rewilding and reintroductions.

Species reintroductions and the other-than-human animals involved in rewilding

Red kites are a bird of prey native to the Palaearctic, although they are now largely confined to Europe. They were essentially extinct in Britain by the late 1800s with the primary reason for their decline being persecution by humans. Wildcats are a small felid native to Eurasia. They have been declared functionally extinct in Britain and, again, persecution is the main reason for their decline. Both the red kite and the wildcat are predators and a large reason for their persecution was their (perceived) conflict with human interests in preying on domestic (particularly game) species. Indeed, the recent reintroduction of the red kite and the current reintroduction of the wildcat are causing considerable consternation among those who are concerned that human interests are being, or may be, compromised as a result of the reintroductions.

These human attitudes to other-than-human animals are interesting and are perhaps particularly pronounced in Britain, where concerted efforts were made to eradicate all species which were judged to threaten human interests, including and especially large predators such as bear (*Ursus arctos*), wolves (*Canis lupus*) and lynx (*Lynx lynx*). There is now, however, a moral imperative to reintroduce extirpated species, especially if humans were the cause of their extirpation (Crowley *et al.*, 2017) and the IUCN (2021) notes that 'the ultimate goal of rewilding is the restoration of functioning native ecosystems complete with *fully occupied trophic levels*' (emphasis added). Moreover, the European Union Habitats Directive (1992) mandates that countries should consider reintroducing missing species where the cause of their decline has been addressed. Reconciling these moral, ecological and political imperatives to reintroduce species with human interests and values is one of the major challenges for rewilding and one of the major themes of this book.

Rewilding both promises and demands 'a paradigm shift in the relationship between humans and nature' (IUCN, 2021), the state and extent of this shift and the resultant relationships is one of the most pressing questions for rewilding. This book examines such questions in relation to rewilding in Britain, with Britain presenting a particularly interesting opportunity to do so; sociocultural attitudes to animals, conservation and hunting in Britain, its intensely cultural landscapes from which many native species are missing (often as a result of deliberate eradication), and the impossibility of many of these species returning without human intervention make Britain highly unusual, offering unique insights into human-animal relationships. Although these relationships are unlikely to be replicated exactly in rewilding in other countries, insights from Britain will certainly have relevance in other contexts.

Human-animal relations and the biopolitics of rewilding

Human-animal-environmental relationships are thus the crux of this book and *biopolitics* (Foucault, 1978) is used to explore human governance of other-than-human animals. For all its claims and aspirations to increase other-than-human agency, rewilding entails considerable 'regulation of', and 'exercise of power over' the other-than-humans involved (Foucault, 1978). Further, like other forms of conservation, while rewilding may have benign intent, this exertion of power in an attempt to order life involves a considerable amount of control and therefore harbours the potential for violence (Foucault, 1978). This is evident in rewilding not only in how biopolitics itself plays out but also in the very categorisation of other-than-human animals which determines which mode of biopolitics is applied to them. Three categories in particular are discussed in this book: pest, feral and hybrid.

Pest is a subjective category and may be applied differently depending on time, place, human attitudes, other-than-human animal population size and many other factors, all of which affect human attitudes to other-than-human animals. Pest is also a pejorative term and reflects a value judgement being made. This value judgement, and being categorised as a pest, is, in and of itself, harmful to the other-than-human animal concerned but can also be dangerous, often making them subject to violent forms of control.

Feral is also a pejorative category and has 'pestilent undertones' (Scasta, 2019); feral carries connotations of being a pest meaning that an other-than-human animal may also be categorised as a pest by 'virtue' of being categorised as feral. Strictly speaking, feral simply refers to a domestic animal which is 'no longer living under human control' (Hill *et al.*, 2022). There are, however, two interesting points which should be noted here. First, there could be said to be 'degrees' of ferality. This spectrum could range from previously domesticated other-than-human animals which are free from human control but are still entangled with human society, living in its margins and exploiting its resources, to those which are living entirely separately from and independently of humans (Palmer and Thomas, 2023). This latter example brings us to the next point – ferality appears to be an inescapable category, condemning other-than-human animals to 'feral lives' in which they are othered, excluded and devalued (Johnston, 2021). According to this perspective, domestication is a one-way street – other-than-human animals can be domesticated, moving from wild to domestic, but domestic animals cannot become wild, being unable to escape their ferality even after generations of 'wild' lives (Palmer, 2016).

Hybridity is, perhaps, a more objective term but it is still fraught since, like ferality, there can be 'degrees' of hybridisation and the very concept of hybridisation rests on the troubled concept of 'species'. For expediency, this book relies on the biological species concept for which the definition of a species depends on their ability to breed and produce fertile offspring. European wildcats and domestic cats challenge this concept immediately, being classified by humans as separate species (*Felis silvestris* and *Felis catus*, respectively) but being capable of interbreeding and producing fertile offspring, thereby introducing the hybrids discussed in this book. This troubling of species makes notions of hybridity somewhat arbitrary as does evolution and 'species' development itself which results in various genetic admixtures. Where this is compounded by recent interbreeding, genetic testing to attempt to differentiate species or determine levels of hybridisation becomes far more subjective than objective. This is further compounded by the fact that, again like feral, hybridity can carry negative connotations of impurity, particularly in species-based conservation (von Essen and Allen, 2016). This is explored further in Chapter 7.

Something should also be said here about the term *other-than-human animals* (OTHAs), which is used in this book to refer to all living beings that are not members of the species *Homo sapiens*. The term other-than-human animal is preferred (as opposed to animal, or non-human animal, for example) to acknowledge that humans are also animals, and that while the *otherness* of other-than-human animals should be recognised (and perhaps even emphasised, something which *OTHA* being an apronym assists with), they should not be considered *less-than* human. If and when the term *animal* appears in this book it is either for a specific purpose, for reasons of expediency, or it is an error, and the term other-than-human animal was probably intended.

Chapter 1: Conservation, rewilding and reintroductions in England and Britain

Chapter 1 explores conservation in England and Britain, focusing on reintroductions and rewilding, and on how rewilding evolved and continues to evolve. Rewilding emerged gradually in Britain, incorporating elements from rewilding in the USA and Europe and focusing on rewilding practice rather than theory. This is perhaps why rewilding in Britain is so diverse and contested, and why the concept of family resemblance is useful when approaching rewilding from the British perspective. This perspective is influenced by Britain's human and physical geographies with Chapter 1 offering a discussion of Britain's 'natural' and 'cultural' landscapes. While, ultimately, it is impossible to separate nature and culture in landscape and to set humans apart from nature, for the purposes of discussion, the chapter introduces a typology of rewilding based on the influence on a landscape (along a spectrum from natural to cultural) and the location of agency (from human to other-than-human). Landscapes can thus be considered ones in which nature or culture is more or less active and in which human or other-than-human agency is more or less dominant. The role of humans in rewilding is fraught given rewilding's ambitions to reduce human intervention in ecosystems and to increase other than human agency, and this tension is particularly evident in Britain. This tension is set up in Chapter 1 and then examined in Chapters 2 and 3, using two examples of 'rewilding' in England as case studies, and a potential solution is identified in Chapter 4.

Chapter 2: 'Rewilding' the Avalon Marshes

The first of these case studies is the Avalon Marshes, an area of wetland in the southwest of England undergoing ecological restoration following peat extraction and intensive farming (i.e., 'major human disturbance' as identified in the IUCN definition of rewilding). Conservation efforts in the Avalon Marshes predate rewilding's popularity as a term in Britain and those involved in the project do not necessarily use the term to describe their activities. Nonetheless, a form of rewilding is taking place in the Avalon Marshes with an increase in biodiversity and an increase in ecological functioning. Interestingly, while there has been a reduction in human intervention in the landscape, elements of the restoration have involved, and indeed still involve, significant human intervention, not least because of the heritage value of the landscape with humans having lived in the area since the Bronze Age. An additional factor is that people still live in the area and, given that the Avalon Marshes are a wetland, significant human intervention occurs in ongoing negotiations with water in an attempt to satisfy the needs of the human and other-than-human animals who dwell in the landscape.

Chapter 3: Wilding the Ennerdale valley

The second case study is the Ennerdale Valley, a water catchment in the Lake District in England's northwest. Like the Avalon Marshes, Wild Ennerdale does not explicitly call itself rewilding, having adopted the epithet 'wilding'. Also like the Avalon Marshes, Wild Ennerdale is undergoing ecological restoration following intensive farming and forestry in the area. Sheep farming is an integral part of the Lake District's culture, and this makes the changes to it that are occurring at Wild Ennerdale a point of tension for its stakeholders, not least because the Lake District itself is so important in the English cultural imagination of landscape. Because sheep farming has contributed to the creation of the Lake District landscape, and because sheep farming involves human intervention in the landscape, the reduction of human intervention inherent in the (re)wilding of the Ennerdale Valley is challenging to stakeholders both for the loss of farming and for the loss of the landscapes associated with it. (Re)wilding in Wild Ennerdale is attempting to negotiate this by maintaining opportunities for farming, albeit extensive cattle farming rather than intensive sheep farming. Indeed, domestic cattle (*Bos taursus*) are performing naturalistic grazing, simultaneously filling the role of wild herbivores acting as disturbance factors within an ecosystem, increasing biodiversity and increasing ecological functioning, while also being part of a productive agricultural system.

Chapter 4: Domesticating rewilding in England and Britain

Chapter 4 draws together the previous three chapters to analyse the state of rewilding in England and Britain. Since the concept of rewilding is set out in Chapter 1 and the practice of rewilding in England is examined through the case studies in Chapters 2 and 3, it is possible to draw conclusions about rewilding in the English context in this chapter. Having observed the way that rewilding is negotiated in the Avalon Marshes and Wild Ennerdale, it is possible to note adaptations to the traits identified as conferring family resemblance to rewilding. For example, the ongoing human *intervention in* landscapes in the Avalon Marshes is reframed as *interaction with* landscape, indicating a shift in approach to the core trait of 'reducing human intervention'. Similarly, in Wild Ennerdale, the deployment of domestic cattle as naturalistic grazers indicates the change in form which rewilding in England is taking, including *functional diversity* as well as, or even in some cases instead of, *biodiversity*. Furthermore, the name Wild Ennerdale illustrates a change in terminology in England (and Britain) from *rewilding* to *wilding*.

Adaptations to the other factors that can confer family resemblance to rewilding are also evident, the Avalon Marshes and Wild Ennerdale (and other English and British rewilding sites) are small as compared to rewilding sites in other parts of the world, suggesting that the ambition to operate at large scale is being moderated to fit the British context. A change in perspective is also evident in relation to increasing ecological functioning; while rewilding in Britain does seek to restore natural processes, this is not always possible and there is an acceptance that these processes will either remain absent or be replaced by human processes. Lastly, with respect to increasing other-than-human agency, this is evident in Britain but to a somewhat limited degree, with the recognition that there is a *biopolitics of rewilding* in which humans govern the lives and deaths of the other-than-human animals involved.

Chapter 5: Let live and let die: The biopolitics of rewilding and (self)-governing cows

Biopolitics is explored in Chapter 5, examining Foucault's 'make live, let die' with respect to rewilding. Biopolitics are a way of governing life – exercising power over living things to regulate and optimise life. This exertion of this power subjects life to 'precise controls and comprehensive regulations' (Foucault, 1978), to the extent that life can be compelled or denied. Many forms of biopolitics are evident in the many different forms of human–animal relations, with scholars having noted them in relation to conservation generally (van Dooren, 2014; Biermann and Anderson, 2017) and rewilding specifically (Lorimer and Driessen, 2013). Chapter 5 builds on this, identifying four biopolitical modes relevant to rewilding: species as expendable objects, animal machines/human proxies, species as analogues and species as self-determining agents. That these biopolitical modes exist indicates that humans are still involved in the governance of the lives and deaths of the other-than-human animals involved in rewilding projects, it also indicates that agency has therefore not been fully restored to these other-than-human animals. This is particularly evident in relation to three of the modes identified (species as expendable objects, animal machines/human proxies, species as analogues) which all conform to Foucault's logic of make live, let die. It might be expected that, *a propos* rewilding, the logics of life and death would be different, and this is the case in the fourth mode identified – species as self-determining agents. Here, the logic is *let* live, let die, indicating that other-than-human animals in this mode have agency over their lives. Chapter 5 illustrates these biopolitical modes, as far as possible, using the example of cattle. Domestic cattle have become the rewilding 'tool of choice' in Britain since they can perform naturalistic grazing (acting as analogues for the extinct Aurochs) and can also be managed through conventional agriculture systems. Because of this, however, cattle involved in rewilding can become liminal animals – existing in the margins between the wild and the domestic simultaneously being both and yet neither.

Chapter 6: Of kites and men: Valuing, categorising and controlling kites in Britain

Red kites are in a rather different situation, as is explored in Chapter 6. The red kite is native to Britain but, in the late 1800s, was almost entirely extinct there. The programme to reintroduce them is discussed in this chapter alongside the changing human–kite relations which led to and resulted from the reintroduction. These human–kite relations are used to explore the way humans value other-than-human animals and the categories we assign them to. Kites can be classified, and valued, as charismatic wildlife or as pests, in which case they are not valued. Interestingly, kites can move between these categories, depending on context, and can also occupy both categories simultaneously. The way kites are categorised and valued goes on to affect the way they are treated, and this is clearly manifested through human feeding of kites. Again, the feeding of kites is used as an opportunity to explore wildlife feeding and conservation feeding more broadly; feeding can be a form of care but can also be a form of violence and control, with some types of feeding becoming a biopolitical tool. The modes of biopolitics discussed in Chapter 5 are applied to kites in this chapter, with kites occupying the modes of species as expendable objects and species as self-determining agents. The mode kites are assigned to is very much linked to whether they are valued or not and whether they are classified as charismatic wildlife or pests.

Chapter 7: The wildness of cats

The discussion of value, categories and biopolitics is continued in Chapter 7 in relation to wildcats. The situation regarding wildcats in Britain is extremely complex and this chapter raises many issues. Wildcats are native to Britain and other parts of Europe but, while they are not at risk in other parts of their range, they are critically endangered in Britain. Indeed, they have been classified as functionally extinct there and there is uncertainty over whether any 'pure' wildcats remain in the wild or whether any free living 'wildcats' in Britain are in fact hybrids. Hybrid is an important category with respect to wildcats. Hybrid cats are the result of breeding between the wildcat and the domestic cat and thus trouble the wild/domestic boundary and frustrate human classification systems. Like the cattle involved in rewilding projects, hybrid cats are liminal animals – both wild and domestic and neither wild nor domestic. Feral cats are similarly problematic both from a classification point of view and from a conservation point of view. Human care for and about these cats is the focus of this chapter and, in discussing this, the chapter also discusses the way humans categorise and value cats and how they control them, including a discussion of the biopolitical modes which wildcats occupy (species as expendable objects, species as analogues and species as self-determining agents). The discussion in this chapter is centred around attempts to reintroduce the wildcat to Britain and this leads to a discussion about types of wildness (genetic, functional, authentic and new) all of which raise further questions around how humans value and categorise other-than-human animals.

Chapter 8: Gods and monsters: The charismatic megafolk of rewilding

Chapter 8 examines the human animals of rewilding – in particular, the *charismatic megafolk*. Charisma makes people and their ideas compelling to others and it is commonly seen in conservationists. The conservation concepts of *charismatic megafauna*, and the less used concept of *charismatic megaflora*, in which fauna and flora are appealing to people, have been extended to humans to discuss the larger than life, charismatic humans involved in conservation generally but rewilding specifically. Charisma and celebrity are closely linked and some of conservation and rewilding's charismatic megafolk are celebrities such as David Attenborough, the late Steve Irwin and Chris Packham. To examine how charismatic megafolk can and do influence rewilding, this chapter identifies the types of charismatic megafolk present in rewilding and the roles they play. These charismatic megafolk are rewilding trendsetters, rewilding soap boxers, guerilla rewilders, and narcissistic and Machiavellian rewilders. As is apparent from these names, charisma has a 'dark side' and can create monsters as well as gods. Thus, while all the charismatic megafolk discussed in this chapter *intend* to advance the cause of rewilding, in the case of guerilla rewilders and narcissistic and Machiavellian rewilders, they may actually be detrimental to rewilding progress because of their behaviour traits. While charismatic individuals are confident and altruistic, have sound judgment and behave responsibly, individuals with dark charisma are arrogant and egoistic have poor judgement and behave irresponsibly. Nonetheless, whichever kind of charisma these megafolk possess, they are highly influential in the discourse and practice of rewilding.

References

Biermann, C. and Anderson, R. (2017) 'Conservation, biopolitics, and the governance of life and death', *Geography Compass*, 11(10), pp. 1–13.
Campbell, T. (2006) '"Bios," immunity, life: The thought of Roberto Esposito', *Diacritics*, 36(2), pp. 2–22.
Crowley, S. L., Hinchliffe, S. and McDonald, R. A. (2017) 'Nonhuman citizens on trial: The ecological politics of a beaver reintroduction', *Environment and Planning A*, 49(8), pp. 1846–1866.
van Dooren, T. (2014) *Flight Ways - Life and Loss at the Edge of Extinction, Nucleic Acids Research*. New York: Columbia University Press.
von Essen, E. and Allen, M. (2016) 'Wild, but not too-wild animals: Challenging Goldilocks standards in rewilding', *Between the Species*, 19(1), pp. 80–108.
European Union (1992) *Council directive on the conservation of natural habitats and of wild fauna and flora*. Brussels: European Commission. Available at: https://eur-lex.europa.eu/legal-content/EN/TXT/?uri=CELEX%3A01992L0043-20130701
Foucault, M. (1978) *The Will to Knowledge*. London: Penguin.
Hill, K., Szydlowski, M., Heaney, S. O. and Busby, D. (2022) 'Uncivilized behaviors: How humans wield 'feral' to assert power (and control) over other species', *Society and Animals*, 1, pp. 1–19.
IUCN CEM (2021) *Rewilding Principles*. Gland: IUCN. Available at: https://iucn.org/sites/default/files/2022-10/principles_of_rewilding_cem_rtg.pdf
JNCC (1996) *Review of Information, Policy and Legislation on Species Translocation*. Peterborough: JNCC. Available at: https://citeseerx.ist.psu.edu/document?repid=rep1&type=pdf&doi=1cb82ca11387e92ea76f5ca64d24e56c786d2669
Johnston, J. J. (2021) 'Imposing ferality: A technopolitical analysis of feral and free-roaming animal classification technologies', *Urban Geography*, 44(10), pp. 2163–2184.
Jørgensen, D. (2015) 'Rethinking rewilding', *Geoforum*, 65, pp. 482–488.
Lorimer, J. and Driessen, C. (2013) 'Bovine biopolitics and the promise of monsters in the rewilding of Heck cattle', *Geoforum*, 48, pp. 249–259.
Martin, A., Fischer, A., McMorran, R. and Smith, M. (2021) 'Taming rewilding - from the ecological to the social: How rewilding discourse in Scotland has come to include people', *Land Use Policy*, 111, pp. 1–12.
Murray, M. (2017) 'Wild pathways of inclusive conservation', *Biological Conservation*, 214, pp. 206–212.
Palmer, C. (2016) 'Climate change, ethics, and the wildness of wild animals' in B. Bovenkerk and J. Keulartz (Eds), *Animal Ethics in the Age of Humans: Blurring Boundaries in Human-animal Relationships*. Cham: Springer, pp. 131–150.
Palmer, A. and Thomas, V. (2023) 'Categorisation of cats: Managing boundary felids in Aotearoa New Zealand and Britain', *People and Nature*, 5, pp. 1539–1551.
Scasta, J. D. (2019) 'Why are humans so emotional about feral horses? A spatiotemporal review of the psycho-ecological evidence with global implications', *Geoforum*, 103, pp. 171–175.
Wittgenstein, L. (1968) *Philosophical Investigations*. London: Blackwell.
Woods, M. (2005) 'Ecological restoration and the renewal of wildness and freedom' in T. Heyd (Ed) *Recognizing the Autonomy of Nature: Theory and Practice*. New York: Columbia University Press, pp. 170–188.

1 Conservation, rewilding and reintroductions in England and Britain

The origin and evolution of rewilding and reintroductions in England and Britain

Rewilding can be traced back to The Wildlands Project in North America in the late 1980s (Jørgensen, 2015; Lorimer et al., 2015). By 1992, Martin wrote that '[t]o behold the Grand Canyon without thoughts of its ancient sloths, goats and condors is to be half-blind' (1992, p. 31). Although Martin (1992) did not use the term rewilding, his reference to absent species and prior ecosystem states touches on rewilding's central themes and the editorial of the journal in which he was writing ('Wild Earth' – the journal of *The Wildlands Project*) does use the term, speaking of the 'living earth rewilding itself' (Foreman, 1992). The term *rewild* had first appeared in print two years earlier in 'Newsweek' (an American weekly news magazine) which spoke of activists vowing 'to take back and 'rewild' one-third of the United States (OED, 2020a). It was not until 2005, however, that the term gained serious traction in the scientific press (Lorimer et al., 2015), and it would be another four years before there was a real increase in its use in the general media (Jepson, 2016).

Rewilding's origin was not a parthenogenesis – there are two widely used conservation terms which precede rewilding and which can be said to contribute to its conception: *wilderness* and *translocation*. Wilderness is perhaps a particularly important concept in the USA, given its association with the wilderness movement and the Wilderness Society, which was founded in the 1930s (Soule and Noss, 1998). In the USA, the term wilderness has distinct connotations of large areas of land from which humans are excluded (Hintz, 2007; Lorimer et al., 2015; Pettorelli et al., 2018), indeed the Wilderness Act defines wilderness as 'an area where the Earth and its community of life are untrammelled by man, where man himself is a visitor who does not remain' (1964, p. 1). What is notable in this definition is that it excludes humans from the Earth's 'community of life', linking the nature/culture debate (discussed later in this chapter) to wilderness, and suggesting that humans and wild nature cannot coexist. Cronon identifies the crux of this debate as being that wilderness embodies 'a dualistic vision in which the human is entirely outside the natural' (1995, p. 79). This dualism, and its more nuanced alternative,

DOI: 10.4324/9781003456490-2

have come to preoccupy the debates and practices of rewilding and these questions will recur throughout this book.

This concept of wilderness evolved, at least in part, from *monumentalism* which saw the creation of some of the USA's national parks to be *national monuments* on the basis of their spectacular natural beauty (Soule and Noss, 1998). In turn, the wilderness movement evolved into *biological conservation*, whereby areas were recognised for their biodiversity and biological importance rather than simply their aesthetic appeal (Soule and Noss, 1998). Over time, biologists (e.g., Frankel and Soule, 1981) realised that in order to maintain this biodiversity, species must be allowed sufficient space to insulate them from the risk of extinction. In regions where wilderness or conservation reserves were not large enough to provide this, the need for connectivity was recognised with 'inter-regional connectivity seen as necessary for providing genetic and demographic rescue and for viability of wide ranging species' (Soule and Noss, 1998, p. 21). This idea then progressed into the cores, corridors and carnivores concept of early rewilding theory in the USA (Soule and Noss, 1998, p. 22), which we will return to later.

In contrast to the relatively subjective *wilderness*, *translocation* is a more technical term. It is used by both the International Union for the Conservation of Nature[1] (IUCN) and the Joint Nature Conservation Committee[2] (JNCC), which define the term as the 'human-mediated movement of living organisms from one area, with release in another' (IUCN, 2013, p. 2) and 'the transfer by human agency of any organism(s) from one place to another' (JNCC, 1996, p. 13). The Oxford English Dictionary (OED, 2020b) gives the first use of the term (in the context of wild animal movement) as an 'Oryx' article from 1962 which opens with its own definition of translocation – 'the transfer of wild animals from one area to another' (Harthoorn, 1962, p. 215). In chronological terms this situates translocation as emerging at a historical midpoint between the wilderness and rewilding movements and its influence on the origins of the rewilding movement is self-evident, especially as Harthoorn goes on to explain that the term 'may be used in the positive sense of introducing animals into an area, for instance to enhance the value of a reserve by re-introducing species that have been exterminated there' (1962, p. 215). Here, Harthoorn (1962) touches on one of the major arguments in favour of rewilding via species reintroduction – that it can be a means of 'restoring function to impoverished ecosystems' (Sandom, Hughes and Macdonald, 2013, p. 336). This theory often rests on the impact of keystone species[3]; if keystone species are absent, then ecosystem processes can be 'missing or dysfunctional' (Prior and Brady, 2017) with many forms of rewilding seeking to address this by reintroducing the species in question.

Given that rewilding has evolved, at least in part, from the American concept of wilderness and that it had its genesis in the USA, rewilding's transition from the USA to other parts of the world, specifically to England and Britain, is pertinent to this book. Rewilding's advent in Britain is more recent than in the USA and is more practical than theoretical (Taylor, 2004;

Brown, McMorran and Price, 2011). This practical emergence can be seen as arising from concerns that existing conservation techniques were failing to arrest biodiversity loss and the perception that an alternative approach was needed (Taylor, 2005; Sandom and Wynne-Jones, 2019). Rewilding in Britain thus emerged gradually as conservation modes shifted from habitat *protection* to habitat *creation*, with an inchoate rewilding network appearing and then being bolstered by the founding of the Wildland Network in 2005[4] (Taylor, 2004). Together with its networking activities, the Wildland Network produced *Beyond Conservation – a Wildland Strategy* (2005), a volume which collated information on rewilding-like projects in Britain and attempted to 'review and address the obstacles to further progress' (Taylor, 2004, p. 2), a clear sign that a move to implement rewilding as a conservation approach was developing.

In 2006, members of the Wildland Network visited the Oostvaardersplassen, a Dutch rewilding project, to learn from the approach to wild land 'management' there, including Frans Vera's theory of grazing ecology. Vera's (2000) theory regarding the effect of grazing on the landscape is credited with disrupting the received wisdom that the climatic climax vegetation in Europe would have been closed-canopy woodland: Vera suggested that it would have been a network of trees and glades created by the combination of ecological succession[5] and the dynamic forces of large herbivores acting as disturbance factors.[6] Vera was developing this theory, including the influential role that certain species play in ecosystems, independently of but contemporaneously with the development of rewilding in the USA. The development of his theory at the same time as similar ideas emerged in the USA led to Vera's work becoming highly influential for rewilding projects in Europe, starting with the Oostvaardersplassen which he was instrumental in establishing (Lorimer and Driessen, 2014). The visit to the Oostvaardersplassen by the Wildland Network ensured that by 2006 rewilding's principles were becoming established in conservation practice in Britain, informed by American and European examples and by American terminology (Fisher, 2004).

Rewilding in Britain moved from academic and conservation circles to wider media and public debates circa 2013 with the publication of *Feral* by George Monbiot (Jørgensen, 2015; Linnell et al., 2015; Carey, 2016; Jepson, 2016; Olwig, 2016; Tanasescu, 2017; DeSilvey and Bartolini, 2018; Wynne-Jones, Strouts and Holmes, 2018; Gammon, 2018; Deary and Warren, 2018; Pettorelli et al., 2018; Sandom and Wynne-Jones, 2019). In *Feral*, Monbiot (2013) discusses rewilding as a *positive environmentalism*, while being highly critical of the state of Britain environment. He is particularly censorious regarding the uplands, describing them as *sheepwrecked* (denuded of flora due to overgrazing by sheep) and suggesting that the lack of flora translates to an absence of fauna, since the foundations of the food web were missing. This description of the uplands, and the proposal of rewilding as a solution, was highly unpopular with farmers who disputed both the diagnosis and the remedy. The book served as a catalyst for public debate on the topic, which

Monbiot continued to engage with in the media, including his own newspaper column (Piesing, 2016). This (ongoing) debate became highly polarised and contentious with scholars noting that *Feral* antagonised stakeholders and that rewilding became toxic as a term and a concept following the book's publication (Jørgensen, 2015; Jepson, 2016; Prior and Brady, 2017; Tanasescu, 2017; Deary and Warren, 2018; Gammon, 2018; Pettorelli et al., 2018; Sandom et al., 2018; Sandom and Wynne-Jones-2019).

While perhaps in the vanguard of popular writing on rewilding, *Feral* is not alone. Since the publication of *Feral*, a plethora of popular books on rewilding have been published, perhaps most notably, certainly from the British perspective, Isabella Tree's books. *Wilding* (2018), an account of the rewilding of the Knepp Estate where Tree lives with her husband, became a 'Sunday Times' best seller in 2019. It was followed by *The Book of Wilding: A Practical Guide to Rewilding, Big and Small* (2023), co-authored by Tree and her husband Charlie Burrell. As a result of their books and the rewilding of their estate Burrell and Tree are prominent in the popular and academic discourse around rewilding in Britain, a discourse which questions what rewilding is and what it means for Britan.

Definitions and discourses of rewilding

When the research on which this book is based was conducted, there was no 'universal' definition of rewilding. This changed in 2021 when the IUCN adopted a lengthy definition of rewilding, quoted here in full. According to the IUCN (2021) rewilding is:

> the process of rebuilding, following major human disturbance, a natural ecosystem by restoring natural processes and the complete or near complete food-web at all trophic levels as a self-sustaining and resilient ecosystem using biota that would have been present had the disturbance not occurred. This will involve a paradigm shift in the relationship between humans and nature. The ultimate goal of rewilding is the restoration of functioning native ecosystems complete with fully occupied trophic levels that are nature-led across a range of landscape scales. Rewilded ecosystems should – where possible – be self-sustaining requiring no or minimum-intervention management (i.e., *natura naturans* or 'nature doing what nature does'), recognising that ecosystems are dynamic and not static.

The fact that this definition is so long and was accompanied by ten 'rewilding principles', reveals the complexity of rewilding and the difficulty of summarising it neatly or succinctly. Indeed, rewilding has been called 'slippery' (Cloyd, 2016), and 'plastic' (Jørgensen, 2015), highlighting its ineffable nature. Despite there now being an IUCN definition of rewilding, debates continue regarding what it *can* mean as much as what it *does* mean. There is, for example, ongoing discourse (and discord) over whether rewilding can, or *should*, involve human intercession (Linnell et al., 2015), whether it is forward or backward looking (Crowley, Hinchliffe and McDonald, 2017; Deary and Warren, 2018) and whether it is toxic (Sandom et al., 2018) and threatening (Carver, 2016) or positive (Jepson, 2016) and has caught the public imagination (Deary and Warren, 2018).

One of the foci of the debate surrounding rewilding centres on its prefix 're', with suggestions that this implies an intent to return to an earlier ecological state (Carver, 2016; Tanasescu, 2017; Deary and Warren, 2018). Such an intent is considered problematic for two reasons. First, it is argued that any baseline[7] is arbitrary and that it is, therefore, difficult if not impossible to justify the selection of one reference point over another (Seddon et al., 2014; Deary and Warren, 2018). Second, critics highlight the impossibility of returning to a previous state due to changing conditions and changes in species (including extinctions) (Carver, 2016; Deary and Warren, 2018). An alternative reading of the prefix 're', however, is that of returning control rather than returning to a specific time period. Corlett's (2016) definition of rewilding as 'returning a managed area back to the wild' could be interpreted in this sense in that control is transferred from 'human management' to 'wild agency'. Debates such as this are compounded by a multitude of similar and qualified terms (e.g., reintroduction, active rewilding,[8] passive rewilding[9]), so that various understandings of rewilding do not always correspond. This is then exacerbated by the way the term is employed in practice.

Since its inception, the term rewilding has been co-opted by a variety of movements, many of which have redefined it in keeping with their specific agendas, with different organisations setting divergent benchmarks[10] and/or adopting disparate approaches while nevertheless using the same term. Three examples from the UK demonstrate how much diversity there can be in a single nation.[11] Knepp Estate in Sussex, England, (described by Tree, 2018 and Tree and Burrell, 2023) is a project involving the conversion of an intensive dairy farm to an extensive beef farm. The estate has also reintroduced other animals as analogues for extirpated species to act as disturbance factors in the newly evolving ecosystem (Knepp Wildland, 2020a, 2020b). By contrast, Glenlude in the Scottish Borders is working to return a former sheep farm and conifer plantation to 'native habitat' (John Muir Trust, 2020). In Ceredigion, Wales, meanwhile, the Summit to Sea/O'r Mynydd i'r Môr project aimed to restore flourishing ecosystems across an area stretching from the Pumlumon massif down (and even into) Cardigan Bay[12] (Newton, 2019). That these three projects all use the term rewilding, and yet have very different practices and aims, illustrates the diversity with which the term is applied.

Another broad distinction in rewilding approaches is between those which focus on species and those which focus on land. The former tend to centre on the reintroduction of keystone species (contributing to the conflation of the terms rewilding and reintroduction), one of the earliest iterations of which was the cores, corridors and carnivores, or *three Cs* concept (Soule and Noss,

1998). This approach focused on the regulatory roles of large predators with core populations of these species connected by a system of corridors (Soule and Noss, 1998). This was followed by the idea of Pleistocene rewilding whereby the notion of reintroduction was extended in terms of species and also time, so that large vertebrates (not only carnivores) were considered for reintroduction, as were species which could act as analogues for those that became extinct at the end of the Pleistocene epoch (Donlan, 2005).[13] The common theme in these species-centred approaches is that the species themselves are considered important and are seen as crucial to the rewilding agenda.

Another interpretation of rewilding places more emphasis on the land rather than the species in it. In particular, *land abandonment* involves a process, whether intentional or unintentional, in which previously managed land is allowed or encouraged to assume a less-cultivated state (Höchtl, Lehringer and Konold, 2005; Navarro and Pereira, 2012). There can, however, be areas of overlap between land abandonment and species reintroduction, particularly in projects which involve 'the reintroduction of keystone species and ecosystem engineers[14] to restore function to impoverished ecosystems' (Sandom, Hughes and Macdonald, 2013). In such cases, while the presence of the animals may be desirable, their species value is secondary to the value of the function they perform in the landscape.

Factors which can confer family resemblance to rewilding

Despite the profusion of interpretations of rewilding, common themes are recognisable, with scholars identifying a 'clustering' of concepts centred on rewilding (e.g. Jørgensen, 2015; Prior and Ward, 2016; Svenning et al., 2016; Prior and Brady, 2017; Tanasescu, 2017; Gammon, 2018; Sandom et al., 2018; Wynne-Jones, Strouts and Holmes, 2018). In light of this (and given that the IUCN definition of rewilding did not exist when the research on which this book is based was started), rather than proposing yet another definition of rewilding to add to the proliferation of existing definitions, the research identified factors which could confer *family resemblance* to rewilding (Wittgenstein, 1968). Family resemblance hinges on the idea that members of a group share different combinations of a collection of common traits and while individuals are different, they possess broad similitude by virtue of the way their traits overlap (Wittgenstein, 1968). Given rewilding's multifarious interpretations and applications, family resemblance is an extremely useful concept to apply to its classification. No factor is individually 'necessary' or 'sufficient' to confer rewilding status, yet combinations of these factors can be considered 'jointly sufficient' for a project to be considered rewilding (Barrenechea and Castillo, 2019). Conservation initiatives then may display different, though overlapping, combinations of these factors such that they possess family resemblance to the broader concept of rewilding. The list of common factors identified as conferring family resemblance to rewilding are: (i) identifying as rewilding, (ii) functioning at large scale, (iii) increasing biodiversity, (iv) increasing ecological functioning, (v) reducing human intervention and (vi) increasing other-than-human agency. A conservation project need not demonstrate all six factors: possessing two or more factors is sufficient for a project to be classified as rewilding.

Definitions and discourses of landscape

Rewilding occurs within human and physical geographies and very much within the context of humans and their activities. While the initial aim of rewilding was perhaps *ecological* restoration, the *eco-social* restoration aspects of rewilding are equally, if not more, important. Rewilding calls for, necessitates, and offers the opportunity for a change in human–animal–environmental interactions. This is highlighted by the IUCN's (2021) definition of, and principles for, rewilding. The definition suggests that rewilding occurs in ecosystems that have experienced 'major human disturbance' (such as peat extraction in the Avalon Marshes and sheep farming in Wild Ennerdale as will be seen in the following chapters) and require a 'paradigm shift in the relationship between humans and nature', with the implication that this shift will lead to a reduction in human intervention in rewilded ecosystems (IUCN, 2021). The shift in the relationship between humans and nature was apparently viewed as crucial by the definition's authors, since it is singled out to be one of the ten rewilding principles. Other principles also refer to the human aspect of rewilding, particularly that it will 'require local engagement and support' (IUCN, 2021). In a less practical way, the principles also note that rewilding should be 'informed by both science and indigenous and local knowledge', alluding to the need to recognise local understandings, values and perspectives of rewilding landscapes. Lastly, and in a slightly less direct way, the principles identify that rewilding must be 'adaptive and dependent on monitoring and feedback' (IUCN, 2021). While the description of this principle does not explicitly refer to rewilding responding and adapting to *sociocultural* feedback, it is essential that it does so in order to be able to operate within *socioecological* systems. England and Britain are unusual, even unique, in their socioecological systems. The dense entanglement of their human and other-than human worlds make them particularly interesting cases with which to explore rewilding and reintroductions: beginning that exploration requires an appreciation of their landscapes.

Landscape has been variously described as 'elusive' (Green, 1996), 'spectral ... yet somehow clear and distinct' (Wylie, 2007, p. 1) and 'an ambiguous and multivalent term that resists a convenient definition' (Zeller, 2007, p. 13). As is evident from these descriptions, the concept, like rewilding, is a problematic one and, again like rewilding, attempts to define it have led to a proliferation of definitions and interpretations. Part of the difficulty is that there are two elements to landscape: it can be an object that can be experienced bodily and a subject that can be perceived cerebrally or emotionally. These two elements are sufficiently related to impinge on and influence each other but sufficiently distinct as to set up a tension between 'the phenomenon itself and our perception of it' (Wylie, 2007, p. 7) and whether a landscape is a lived in world or a looked at scene. This is highly relevant to rewilding for two reasons. First, it determines the extent to which humans can inhabit landscape and be involved in its creation, something which is fraught for wilderness (given the connotation of wildness excluding people discussed earlier) and still problematic for rewilding with its attempts to reduce or remove human influence and artefacts. Second, and highly relatedly, because landscape is heavily dependent on our perception of it, different people can look at the same landscape and see very different things: a 'chocolate box' countryside to visit, a farm or rural community where people live and work, or a *terra nullius* to rewild.

Different interpretations regarding the term landscape in the English language extend to its etymology. Its origins can be traced as far back as Anglo-Saxon in which the notion of 'land' as a 'space with boundaries' combined with the idea of 'scapes' being 'compositions of similar objects', evolving to become 'the patterns and processes characterising a specifically circumscribed tract of country' (Green, 1996, p. 12). Others suggest that the term arrived much more recently, in the 16th century, as a loanword from the German 'landschaft' (Zeller, 2007), which originally had connotations of 'political community' (Wylie, 2007). In this sense, landscape means a 'judicially defined polity, not a spatially defined area' (Olwig, 2002, p. 17), with landscape being the 'expression of the practices of habitation through which the habitus of place is generated and laid down as custom and law upon the physical fabric of the land' (Olwig, 2002, p. 226). It is this broader interpretation of the term that enables us to approach landscape from a physical, environmental perspective or from a cultural, political and legal point of view (Wylie, 2007).

Meanwhile, the development of new techniques to afford perspective to paintings, and the advent of landscape painting in the 15th century, saw the meaning of landscape gradually shift from that of 'common *places* to scenic *spaces*' (Olwig, 2002, emphasis in original). By the century the dominant understanding of the term was that of the scenic definition (Wylie, 2007) leading to the idea that landscape was something to be viewed. This implied 'separation and observation' (Williams, 1975), setting 'us at a distance', turning us into 'detached spectators', and the world into 'distant scenery to be visually observed' (Wylie, 2007, p. 3). This attitude has resulted in interpretations of landscape which assess it largely on its (natural) aesthetic value, but the landscape is 'more than scenery. It is a place for social and economic activity' (Sheail, 2007, p. 324).

This recognition of the multiple aspects that contribute to landscape sets up an important debate regarding its natural and cultural elements and is acknowledged, to a greater or lesser extent, in various definitions. For example, Cosgrove extends Green's definition to make explicit the fact that Green's

(1996) 'patterns and process' are 'the integration of natural and human phenomena over a delimited portion of the Earth's surface' (Cosgrove, 1984, p. 9). This is useful in terms of viewing a landscape holistically, a concept that is highly pertinent in the discourse surrounding rewilding given the human and other-than-human factors which contribute to shaping landscapes. Similarly, for Wylie, landscapes are 'the product of interactions between sets of natural conditions and sets of cultural practices' or, when in a more culturally influenced state, the 'expressions of human responses to and modifications of natural environments over long periods', an interpretation which considers 'the holistic/ecological perspective of nature and humans interacting within the physical space of the landscape unit' (2007, p. 9). The European Landscape Convention's definition of landscape as 'an area, as perceived by people, whose character is the result of the action and interaction of natural and/or human factors' (CELC, 2018) is a similarly inclusive definition of nature and culture in landscape. It acknowledges landscape as something to be in and as something to be viewed, and highlights the importance of human perception in its creation, something which is particularly important with regard to the English landscape.

Landscape in the English context

While its definition is fraught, it is broadly accepted that understandings of landscape are highly location specific, and the term has unique connotations in Western, European, British and English contexts (Green, 1996; Matless, 1998; Adams, 2003; Wylie, 2007). The 'nostalgic' agenda of landscape studies in England was set by W.G. Hoskins in the 1950s (Wylie, 2007). Hoskins' claimed that 'since the year 1914, every single change in the English landscape has either uglified it, or destroyed its meaning, or both' (1955, p. 231). 'Uglification' and attribution (or destruction) of meaning are of course both highly subjective. Uglification relates primarily to aesthetic qualities, which are almost exclusively responsible for contributing meaning when we approach landscape as something to be viewed. When encountering landscape as a physical entity, however, there are multiple factors that can constitute meaning, including aesthetics but also culture, ecology, economy, identity (individual, local and national), nature, history, politics and society. Furthermore, several scholars argue that landscape should be viewed as a process (rather than a product) and that its meaning is therefore being (re) constructed constantly (e.g., Mitchell, 1994, 1996; Green, 1996; Schein, 1997; Cloke and Jones, 2001; Wylie, 2007). Following this logic, Hoskins' (1955) assertion that change can destroy a landscape's meaning is oxymoronic: change may alter meaning but cannot destroy it. Hoskins' (1955) argument holds only if we accept his idea of a landscape as capable of being completed, as if the English landscape were, at the time of his writing in 1955, the end product of the historical process he describes in his book, rather than another stage in its evolution. What Hoskins (1955) is bemoaning is the loss of a specific, pre-1914 landscape, and the landscape which he is romanticising has inevitably become historic with the passage of time.

Nostalgia is, nonetheless, extremely important in relation to the construction of landscape, particularly rural landscapes; there is a strong sense of this nostalgia in the public imagination which contributes significantly to popular ideas of landscape, and national and personal identity. For example, rural areas have been identified as 'idyllic places of peace, as repositories of national identity' (Shucksmith, 2018, p. 297) with rural idylls, in particular, forming part of the English national identity (Garrard, 2020). In the English context, 'rural idylls' relate to ideas of a beautiful, pre-war landscape (before it was uglified) and both contribute to and reinforce the 'long-standing and deep-rooted English discourse in which a certain rural idyll is represented as a source of aesthetic, social and ecological harmony' (Wylie, 2007, p. 34). The public imagination of this rural idyll is very strongly linked to Romanticism and harks back to the 1800s when the English landscape, as it is nostalgically thought of, emerged, or rather was developed.

The Romantic Movement was at its height in Europe in the early nineteenth century: it centred on the notion that 'deep thinking' also required 'deep feeling' and hence embraced aesthetic as well as rational ideals (Morrow, 2011). Romantic perceptions and depictions of nature were particularly prevalent in landscape painting which, in addition to affording new ways of *showing* landscape, provided new ways of *viewing* it, having an important influence in culturally conditioning ideas of landscape (Morrow, 2011). Thus, when Romanticism developed landscape painting in the pastoral, picturesque and sublime styles, these styles contributed significantly to the discourse surrounding landscape. The pastoral style in particular tamed the natural environment from the 'classic' wilderness ('something to be feared, an area of waste and desolation inhabited by wild animals, savages, and perhaps even supernatural evil' (Light, 1995, p. 195)), to the 'romantic' landscape, contributing to English ideas about the 'proper forms of landscape and Englishness (neat fields, well-nucleated towns and villages, physically upright citizens)' (Wylie, 2007, p. 118). This view persists in the English national consciousness with the 'chocolate-box' perception of what constitutes 'beautiful landscape' (Short, 2002). Indeed, scholars argue that the way rural existence was 'ennobled' by landscape artists in the Romantic Movement such as John Constable and J.M.W. Turner is still clung to in England and is the root of English resistance to landscape change (Short, 2002; Thompson, 2010).

As important as painting in English notions of landscape, is landscape (or nature) poetry, particularly, and of particular relevance to this book, that of the 'Lake Poets'. The Lake Poets (William Wordsworth, Samuel Taylor Coleridge and Robert Southey) were writing during the Romantic Movement and, like Romantic landscape painting, their poetry (perhaps most notably Wordsworth's) has heavily influenced the public imagination regarding the English landscape, particularly the Lake District landscape (Thompson, 2010, p. 87). Like landscape art, landscape poetry changed the public perception of land as something wild, savage and dangerous to something ordered, peaceful and tranquil (Thompson, 2010). Significantly, Wordsworth's poetry is described as not simply nature poetry but as being 'about the organic relationship between human beings and the natural world' (Thompson, 2010, p. 87), presaging debates over the way humans and nature are intertwined in landscape, and the particular significance of this in England.

No discussion of landscape, particularly in the English context can be complete without reference to landscape gardening. Landscape gardening emerged in the 18th century and designed gardens to represent an idealised version of nature (Bouts, 2010). The work of key figures in landscape gardening (e.g., Capability Brown and Humphrey Repton[15]) was so influential that the English landscape park has indeed come to be seen as *the* English landscape, or at least the *ideal* English landscape, so that it is not untrue to say that 'the English landscape was invented by gardeners imitating foreign painters who were evoking classical authors' (Stoppard, 1993, p. 34). Like painting and poetry, landscape design was part of the Romantic Movement: it idealised an intensely curated landscape that presented a simulacrum of nature and which came to be accepted as real. This entanglement of nature and culture in landscape is the subject of an ongoing debate which is given new relevance in light of rewilding.

Nature and culture in landscape

Like nature, natural landscapes are perhaps more easily defined by what they *are not* than what they *are*; a natural landscape is one which has *not* been modified or impacted by humans, with Hoskins (1955) calling them those 'untouched by man'. Like *wilderness*, however, this renders natural landscapes extremely vulnerable. If landscapes are considered to be exclusively natural entities, 'with no human mark' (White, 1995, p. 175), then any human influence must be viewed as binary rather than graded. Thus, if natural landscapes are only those which have not been affected by humans, then a natural landscape which has been subject to *any* human influence immediately ceases to be so. The term 'cultural landscape' is therefore suggested to describe the way natural landscapes are 'shaped' and 'altered' by humans, recognising that this process has been occurring for millennia (Alagona, 2004; Zeller, 2007; Gillson, Ladle and Araújo, 2011). Even in 1955, Hoskins noted the scarcity of natural landscapes, writing that there remain 'not many places where one can feel with complete assurance that this is exactly as the first inhabitants saw it in "the freshness of the early world"' (1955, p. 17). Seventy years later, and with the designation of the Anthropocene to acknowledge the impact that humans are having on the world, there are surely even fewer, if any, such places; human influence is now so far reaching that it is no longer possible, if it ever were, to distinguish between natural and cultural landscapes – all landscapes are therefore, to some extent, cultural landscapes (Head, 2015). This argument is advanced particularly strongly in the case of

Europe (where there is a 'deep, complex, and ancient intertwining of nature and culture' (Linnell et al., 2015, p. 984) and England where it is suggested that landscapes are the product of years of human activity to the extent that even those which look, ostensibly, natural have been shaped by human actions and are therefore cultural (Carver, 2007; Deary and Warren, 2018). Interestingly, the counter argument also stems from the nature/culture debate: if humans are part of nature, their impact on the landscape can be viewed as natural rather than cultural, therefore human shaping of landscape does not necessarily render it unnatural (Alagona, 2004; Carver, 2007). Both these positions blur, or arguably erase, any possible distinction between natural and cultural landscapes. Either all landscapes are natural or all landscapes are cultural, meaning that the terms 'natural landscape' and 'cultural landscape' are redundant. For this reason, some scholars state that landscape is 'impossible to place on either side of a dualism of nature and culture' (Matless, 2003, p. 231), calling it a 'quasi-object'.[16]

In light of such difficulties, current attempts to define landscape have largely abandoned efforts to separate natural and cultural landscapes, with definitions such as Ingold's overlaying one with the other as 'discursive worlds of culturally constructed significance, laid out upon the substrate of a … physical terrain' (2000, p. 172). Even attempts such as this are, however, accused of 'assuming and reproducing a duality of cultural mind and physical nature and in doing so drawing a line between human and natural worlds' (Wylie, 2007, pp. 155–6). Ingold acknowledges this when he says that 'to suggest that human beings inhabit discursive worlds of culturally constructed significance is to imply that they have already taken a step outside the world of nature' (2000, p. 14): it seems that attempts to acknowledge human involvement in, and contribution to, landscape can do more to suggest that humans are *apart from* nature rather than *a part of* it.

The nature/culture divide

The question regarding whether humans are a part of or apart from nature is a divisive one, with arguments lying anywhere on a spectrum from 'humans are fundamentally a part of nature' (Noss, 1994, p. 188) and 'humans are natural and therefore everything we do may also be considered natural to some extent' (Carver, 2007), to humans are 'entirely outside the natural' (Cronon, 1995) with the 'stain of civilization' tainting 'pristine wilderness' (Alagona, 2004).

Humans as apart from nature

Scholars seeking to 'draw a line' between humans and nature are compelled to pinpoint the moment at which humans diverged from nature. This point is usually proposed as having occurred when humans ceased to be hunter-gatherers and began cultivating crops and domesticating animals (i.e., the

Neolithic period) (Carver, 2007; DeMello, 2012). Importantly, the resultant separation is seen as being due to the changing quality of the relationship between humans and nature: the hunter-gatherer method of food acquisition entailed 'an intimate *interaction* with nature' while farming systems 'entail *intervention* with nature' (DeMello, 2012, p. 37, emphasis added). Following this logic, the development of agriculture was the point at which humans began to control animals and nature and therefore transcend them (DeMello, 2012, see also Carver who sees 'removing post-hunter-gatherer humans and their effect on landscape from the definition of natural as probably the most acceptable solution' (2007, p. 268) in resolving the nature/culture debate). In essence then, *pre*-hunter-gatherer humans can be considered natural, or at least sufficiently natural to be allowed to remain in a 'natural' landscape. While this distinction can be argued to have a temporal basis (circa the Neolithic period), it is also activity based – hunting and gathering is natural while farming is not. This is problematic for two reasons. First, other scholars suggest that even the actions of Mesolithic hunter-gatherers can be recognised as separating them from nature (Hodder et al., 2009). Secondly, relating the shift of humans from within to outside nature to an activity (farming) might imply that modern day hunter-gatherer societies are a part of nature in a way that other human societies are not, and even that the transition from part of, to apart from nature is still occurring in those societies (Wylie, 2007).

A second suggestion for the segregation of nature and culture is the distinction between archaic (i.e., traditional) and modern work, in which archaic work is viewed as *in*structive (i.e., a means by which to know nature) and modern work is viewed as inherently *de*structive to nature (White, 1995). This argument allows for pre-modern work within nature but considers modern work to be outside of, or in opposition to, nature, although it also notes that archaic work is 'vanishing' and 'romanticised' (White, 1995). This is reminiscent of Hoskins' (1955) nostalgia for disappearing landscapes and suggests that it is the very fact that these systems are disappearing which gives them their value. This theory is supported by the existence of initiatives which seek to preserve certain (created) landscapes, or 'semi-natural habitats[17]' (i.e., those in which a pre-technological, as opposed to a post-technological, human presence is appreciable) due to their perceived heritage value and threats to their existence (for a discussion of this, see Harrison, 2016; van der Zanden et al., 2017).

Humans as a part of nature

The counter position, that humans are part of nature, argues that humans and nature are inextricably intertwined with an 'entangled organic and social history' that cannot be separated (Helmreich, 2005, p. 124). This argument can be made from three different standpoints: (i) that humans were, are and always will be natural irrespective of their interventions in nature and advances in their technology; (ii) that humans were initially part of nature, diverged from it, and have now re-merged with it due to the extensive and pervasive effects of previous and current technologies; or (iii) that nature and culture, and (therefore) the divide between them are purely human constructions.

The first argument is descriptive rather than explanatory, with scholars simply asserting that humans are 'an integral part of, not apart from, nature' (Seddon et al., 2014), 'all equally part of nature' (Helmreich, 2005, p. 124) and that 'to be human is to be part of Nature' (Saunders, 2016, p. 49). This argument considers humans as 'integral members of the ecological community, one member of an inter-reliant ecological system' (Brown, McMorran and Price, 2011, p. 293) and can be extended to take a 'Gaia-like'[18] view wherein humans as a 'species take on the role of managing the planet as a whole' (Bowker, 2000, p. 644). Another facet of this argument addresses the issue of modern technology as separating humans from nature by saying that as technology is the work of humans, who are natural organisms, technology is therefore also natural and thus 'not alien to nature, but integral to it' (Nye, 2000, p. 10). This argument can be followed to its apparently logical conclusion by saying that 'once one decides that human beings, including their 'works', are part of Nature, one quickly slides down the logical slope to the conclusion that, ergo, everything is Nature and hence nothing is. So it becomes a rather redundant notion' (Saunders, 2016, p. 53). This is a similar argument to the one made regarding 'natural' and 'cultural' landscapes and the impracticality of attempting to distinguish between the two.

The second argument suggests that the human–nature relationship has come full circle and that, assuming humans did diverge from nature through their actions, these same actions have precipitated their re-convergence since human influence is now so extensive that it is impossible for any part of the globe to be untouched by it. This global reach of human influence is most often illustrated by reference to anthropogenic climate change (McKibben, 1990; Cronon, 1995; Cassidy, 2012) and it is suggested that since this is having planetary-scale effects it is not possible for any species to remain unaffected by it (Cassidy, 2012). Given that this argument was first put forward as much as 35 years ago (e.g., McKibben, 1990), if we accept that it was the case then, it must be even more so now with the recognition of the Anthropocene which 'surely disrupts any lingering notion that we can think of the environmental and social realms as separate and separable' (Head, 2015, p. 318). This refutation of the possibility of the continued existence of 'pure' nature (if it ever existed) brings humans and nature back into the same fold, re-entwining them after a (debatable) period of separation.

The third argument asserts that divide between nature and culture is a human construction which is socially and historically contingent – humans and their practices are considered part of or apart from nature depending on their social and historical context (DeMello, 2012). Similarly, elements of nature are considered wild or domestic depending on which side of the (literal and metaphorical) fence they are on. The European rabbit (*Oryctolagus cuniculus*) is an excellent example of this since the 'wild' rabbit is genetically indistinguishable from its 'domestic' counterparts (DeMello, 2012). That social convention allows humans, and their artefacts, and indeed other-than-human species, to cross and re-cross the 'boundary' between the natural and the cultural emphasising how entangled organic and social history is and therefore argues for the inseparability of humans and nature.

To a certain extent, this issue of humans as either part of or apart from nature is irreconcilable and this book cannot and does not hope to resolve it. Indeed, as with landscape, many scholars are now rejecting the idea that they are separate or separable, given that the boundary is 'blurry, diverse, and dynamic' (Linnell et al., 2015) and also 'porous and permeable' (DeSilvey and Bartolini, 2018). As a result they are refusing 'the boundaries that cordon nature from culture' (Tsing, 2012) and suggesting alternative concepts such as 'hybrid landscapes' (Fiege, 1999) 'nature-cultures' (Latour, 1993; Haraway, 2003), 'biocultures' (Maffi and Woodley, 2010; Linnell et al., 2015; Fisher and Parfitt, 2016), 'human–natural systems' (Merckx and Pereira, 2015) and 'social–ecological systems' (Pettorelli et al., 2018) all of which are, arguably, valuable and should therefore be valued in their own right. These chimeric systems are extremely significant with regard to rewilding and reintroductions and will be returned to later in this book. In the meantime, they lead on to another important question with respect to rewilding – the role of humans in nature.

The role of humans in nature

Questions regarding the role of humans in nature can be illustrated through the example of reintroductions. As will be discussed in more detail later in this book, humans play a significant role in (re)introducing other-than-human species to old/new areas. When this is done unintentionally, however, it leads to a species being classified as an 'alien' (Tree, 2018). This delegitimises or denies the role of humans as vectors in a functioning ecosystem (Tree, 2018); if introductions carried out unintentionally by humans are classed as aliens, the implication is that we do not consider ourselves a valid member of the ecological community.

Disavowing the role of humans in nature is problematic in relation to rewilding since it has been accused of being conducted in a way that 'disavows human history and finds value only in historical ecologies prior to human habitation' (Jørgensen, 2015, p. 6) and of perpetuating a discourse which 'separates humans from nature and seeks (explicitly or implicitly) to erase human history by valorising the wild without people' (Deary and Warren, 2018, p. 468). This is evident in rewilding's ambition to reduce human intervention and increase other-than-human agency. Arguably, by doing so, rewilding devalues human involvement in ecosystems and privileges other-than-human processes over human ones, casting other-than-human processes as inherently good and human processes as inherently bad. This can be disenfranchising, not only for those who live and work in or near rewilding areas but for those who value human processes and the role of humans in the landscape;

> I *don't* think that the simple fact of being human is inherently worse than the simple fact of being non-human (i.e., wild) and I find myself increasingly alienated by a conservation narrative which lauds the wild and demeans the human, *per se*
>
> (Saunders, 2016, p. 53, emphasis in original)

From this perspective, something which is achieved via a 'wild' process is not necessarily better than the same result achieved by a human process. Likewise, seeking wildness purely for its own sake is a vain pursuit, it must achieve something *more* than the 'equivalent' human process to be valuable. Conversely, if a human process can offer something over and above that which a wild process can achieve, it may be preferable and would, according to Saunders (2016) offer a 'better' outcome.

In spite of rewilding's ambition to reduce human intervention, some scholars claim that rewilding can avoid reinforcing the nature/culture duality and assert that wildness (as opposed to wilderness) *admits* humans and *permits* their presence in the landscape. Indeed, scholars suggest that

> rewilding acknowledges that humans are an integral part of, not apart from, nature and recasts the retrospective goals of restoring "wilderness" as future-oriented visions of creating "wildness" in which ecological processes are managed within landscapes shared by humans and wildlife'
>
> (Seddon et al., 2014, p. 411)

There is a tension between these apparently contradictory positions regarding rewilding – the goal of reducing human intervention and the aspiration that humans and wildlife can co-exist. This tension will be explored in the two rewilding case studies discussed in the following chapters, and in relation to rewilding in England and Britain more broadly in Chapter 4.

Typology of rewilding

Based on the discussion surrounding agency and landscape in relation to rewilding, in particular the roles that humans and other-than-humans play, and that we might consider landscape to exist on a spectrum of more to less cultural influence, this chapter presents a typology of rewilding (Figure 1.1). The typology is useful in acknowledging the spectrum of rewilding, classifying it according to landscape influence (from cultural to natural, i.e., degree of 'naturalness') and location of agency (from human to other-than-human, i.e., the degree of wildness). According to this typology, rewilding can be

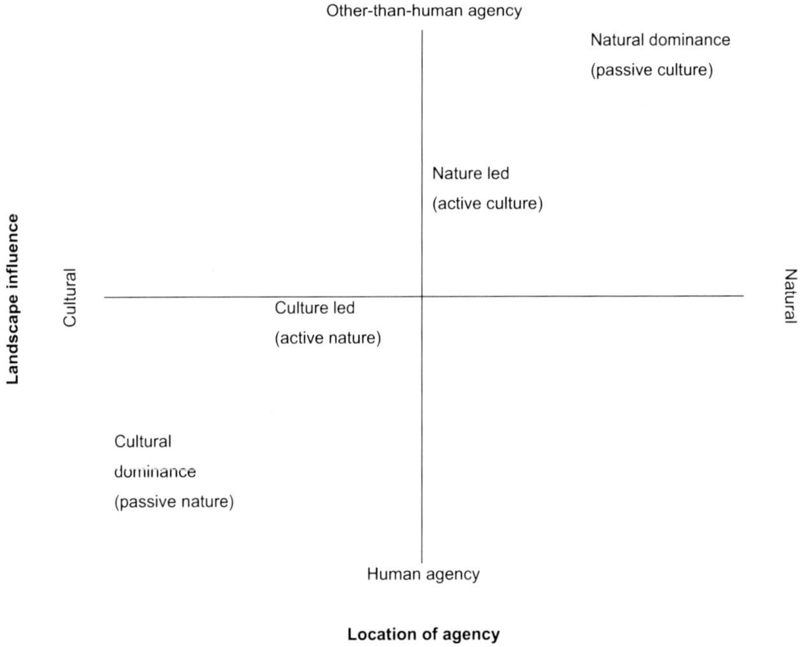

Type	Description	
Natural dominance (passive culture)	Allows natural, ecological processes to resume/continue without human intervention. Can include land abandonment or the deliberate exclusion of humans/human activities (i.e., there is no human intervention).	High
Nature led (active culture)	Encourages natural, ecological processes to resume/continue with human intervention to facilitate, accelerate or moderate this process (i.e., there is low human intervention).	Level of rewilding
Culture led (active nature)	Compels/forces natural, ecological processes to resume/continue via human intervention possibly involving the removal of previous signs of human intervention. Levels of human intervention may be an interim, remedial, restorative step during which more significant interventions occur and then cease, allowing nature to become (more) active, or they may be planned to continue throughout the project (i.e., there is moderate human intervention).	
Cultural dominance (passive nature)	Permits some natural, ecological processes to resume/continue within the bounds of culturally dominated land use. This may not strictly be rewilding but may also include landscape preservation (as opposed to conservation) usually as a result of culturally or historically significant landscapes or structures (i.e., there is high human intervention).	Low

Figure 1.1 Typology of rewilding according to landscape influence and location of agency.

classified as: cultural dominance (passive nature), culture led (active nature), nature led (active culture) and natural dominance (passive culture). Each axis represents a sliding scale rather than binary poles and, while as discussed above, the distinction between natural and cultural landscapes is highly problematic, it is still relevant since rewilding's stakeholders are inclined to perceive landscape as more or less natural or more or less cultural depending on the level of human intervention and the presence of human artefacts.

It should be remembered here that human intervention can be overt or covert: human intervention in rewilding projects often seeks to conceal itself, and/or to undo the work of previous interventions ('using the human touch to erase the human touch' (Hall, 2014)), while human intervention in other types of landscape use (e.g., farming or even conventional conservation) tends to celebrate its effects. Indeed, depending on the extent to which it is accepted that human intervention endures in landscape (and, as we will see later, in other-than-human animals), even apparently 'natural' landscapes (or 'wild' in the case of other-than-human animals) may have been fostered by human activity which is now obscured and are therefore, arguably, 'cultural' (or 'feral' in the case of other-than-human animals).

Reconsidering rewilding

This chapter has traced the origin and development of rewilding in England and Britain as it has travelled and experienced conceptual stretching from its origins in the USA and Europe. While rewilding in Britain, indeed internationally, is a novel approach, it has emerged from a long history of conservation theory and practice, in particular ideas of wilderness and practices of translocation (something which is closely related to reintroductions to which we will turn later in this book). Wilderness and translocations raise questions about the place for and role of humans in nature and conservation, questions which have been transferred to, and even intensified within, the rewilding movement. This is particularly relevant in Britain; with its long history of human habitation and the size and density of its human population today, Britain's landscapes are shaped by an ongoing negotiation between ecological processes and human activity. Rewilding seeks to shift the balance of power within this negotiation, reducing human intervention and increasing other-than-human agency. Rewilding thus has profound implications in landscapes such as Britain's, and this gives rise to the frictions and tensions which are explored in this book. By seeking to withdraw human control, rewilding highlights the ubiquity of human influence, compelling us to reconsider what is meant by 'natural' and 'wild', in the Anthropocene.

This chapter has established two frameworks which help us do this. First, a set of core traits which confer *family resemblance* to rewilding. The concept of family resemblance enables a holistic, inclusive discussion of rewilding, in its many and varied forms, rather than restricting the discourse to rigid definitions and strict constraints. The traits which can confer family resemble to rewilding include *reducing human intervention* and *increasing other-than-human agency*, and also, *identifying as rewilding, functioning at large scale, increasing biodiversity* and *increasing ecological functioning*. Second, a typology of rewilding is introduced which recognises the influence on landscape ('natural' or 'cultural') and the location of agency within it (human or other-than-human). These frameworks provide a foundation for this book's exploration of rewilding and reintroductions in England and Britain and how they reflect and rework the physical and sociocultural constraints within which they operate.

Notes

1 Established in 1948, the IUCN is an environmental network comprised of government and non-government organisations providing advice on conservation, specialising in species survival, environmental law, protected areas, social and economic policy, ecosystem management and education and communication (IUCN, 2019).
2 Originally established in 1990 under the Environmental Protection Act, the JNCC was reformed by the Natural Environment and Rural Communities Act, 2006. It is a public body which advises the British government (and the devolved administrations of Northern Ireland, Scotland and Wales) regarding national and international nature conservation (JNCC, 2015).
3 Keystone species are those which 'in proportion to their biomass have a disproportionately large impact in an ecosystem' (Sandom et al., 2013, p. 431).
4 The Wildland Network is now defunct, having essentially been replaced by the Wildland Research Institute https://wildlandresearch.org/.
5 Ecological succession is the progressive change in structure and composition of an ecosystem over time towards a climax community (Vera, 2000).
6 A disturbance factor is a stochastic element or agent which disrupts, and therefore drives change within, an ecosystem, for example, activity by humans or other-than-human animals (e.g., grazing, browsing and poaching by large herbivores), disease, fire, flood, natural disasters, weather, etc. (Hodder et al., 2009). In the case of cattle, or other large herbivores, their grazing drives 'cyclical vegetation turnover' (Hodder et al., 2005).
7 Baselines are historical reference points which some advocates suggest should be used as ecosystem targets for rewilding (Toledo, Agudelo and Bentley, 2011; Sandom et al., 2013; Seddon et al., 2014; Sandom and Macdonald, 2015; Pettorelli et al., 2018).
8 Active rewilding is an umbrella term used by, for example, Carver (2016) and Sandom et al. (2018) to describe a range of rewilding approaches all of which involve human intervention to a lesser or, more usually, greater extent.
9 While passive rewilding can also be an umbrella term (again employed by Carver (2016) and Sandom et al. (2018)) it has been specifically described by Navarro and Pereira as the 'passive management of ecological succession with the goal of restoring natural ecosystem processes and reducing human control of landscapes' (2012, p. 10).
10 Benchmarks, like baselines, are often used as reference points or targets in rewilding. Unlike baselines, however, benchmarks offer the possibility of being forward looking rather than purely regressive and therefore allow for the creation of novel ecosystems (Hodder et al., 2009; Keulartz, 2009; Toledo, Agudelo and Bentley, 2011; Pettorelli et al., 2018). Benchmarks and baselines, are, however equally

problematic with regard to the rewilding approaches which reject any kinds of goals or targets.

11 Differences can become even more pronounced between continents, for example, the reintroduction of wolves to Yellowstone National Park in the USA is often held up as the quintessential rewilding project and yet bears little or no resemble to the examples given here.

12 It should be noted that the Summit to Sea/O'r Mynydd i'r Môr project was embroiled in significant controversy and underwent major refocussing from being branded as a rewilding project associated with Rewilding Britain to disassociating itself from both the term rewilding and from Rewilding Britain. Nevertheless, it remains an example of landscape scale ecological restoration and still serves as an example of one of the many ways in which rewilding has been interpreted in recent times.

13 The Pleistocene was considered a suitable, and significant, benchmark for rewilding because humans were implicated in its mega-fauna extinctions (Sandom et al., 2013). This identification of human impact on the environment as somehow 'inappropriate' marks a natural/cultural boundary in which humans are seen as *acting on nature* rather than as *acting from within nature* (White, 1995; DeMello, 2012).

14 Ecosystem engineers are 'organisms that demonstrably modify the structure of their habitats' (Wright, Jones and Flecker, 2002).

15 It is notable that the Knepp Estate has a park designed by Humphrey Repton (Knepp, 2020b).

16 The term quasi-object comes from Latour (1993): he uses it to refer to objects that are not quite social and not quite natural.

17 The European Investment Bank defines semi-natural habitats as 'ecological assemblages that have been substantially modified in their composition, balance or function by human activities. They may have evolved through traditional agricultural, pastoral or other human activities and depend on their continuation to retain their characteristic composition, structure and function' (2018, p. 26). This definition is extremely useful for the nature/culture debate since its use of the term 'traditional' implies a distinction between semi-natural habitats and those which have 'evolved' through technologically (i.e., non-traditional) mediated human activities.

18 The Gaia hypothesis postulates that all elements on Earth combine synergistically to form a self-regulating system. According to the hypothesis, humans are not only part of this system but constitute an important part of it, providing a consciousness of and for the system as a whole (Lovelock, 1979).

References

Adams, W. M. (2003) *Future Nature*. London: Earthscan Publications Limited.

Alagona, P. (2004) 'Biography of a "Feathered Pig": The California condor conservation controversy', *Journal of the History of Biology*, 37(3), pp. 557–583.

Barrenechea, R. and Castillo, I. (2019) 'The many roads to Rome: family resemblance concepts in the social sciences', *Quality and Quantity*, 53, pp. 107–130.

Bouts, E. (2010) *Illustrated History of Landscape Design*. Hoboken: Wiley.

Bowker, G. C. (2000) 'Biodiversity data diversity', *Social Studies of Science*, 30(5), pp. 643–683.

Brown, C., McMorran, R. and Price, M. F. (2011) 'Rewilding - A new paradigm for nature conservation in Scotland?', *Scottish Geographical Journal*, 127(4), pp. 288–314.

Carey, J. (2016) 'Rewilding', *Proceedings of the National Academy of Sciences*, 113(4), pp. 806–808.

Carver, S. (2007) 'Rewilding in England and Wales: A review of recent developments, issues, and concerns' in A. Watson, J. Sproull and L. Dean (Eds) *Science and Stewardship to Protect and Sustain Wilderness Values*. Anchorage: U.S. Department of Agriculture, Forest Service, pp. 267–272. Available at: https://www.fs.usda.gov/treesearch/pubs/31040

Carver, S. (2016) 'Rewilding… conservation and conflict', *Ecos*, 37(2), pp. 2–10.

Cassidy, R. (2012) 'Lives with others: Climate change and human-animal relations', *Annual Review of Anthropology*, 41, pp. 21–36.

CELC (2018) *Glossary of the Council of Europe Information System of the European Landscape Convention*. Strasbourg: Council of Europe. Available at: http://www.coe.int/t/dg4/cultureheritage/heritage/Landscape/GlossaryL6_en.pdf

Cloke, P. and Jones, O. (2001) 'Dwelling, place, and landscape: An orchard in Somerset', *Environment and Planning*, (33), pp. 649–666.

Cloyd, A. A. (2016) 'Reimagining rewilding: A response to Jørgensen, Prior, and Ward', *Geoforum*. Elsevier Ltd, 76, pp. 59–62.

Corlett, R. T. (2016) 'Restoration, reintroduction, and rewilding in a changing world', *Trends in Ecology and Evolution*, 31(6), pp. 453–462.

Cosgrove, D. E. (1984) *Social Formation and Symbolic Landscape*. London: Croom Helm.

Cronon, W. (1995) 'The trouble with wilderness; or, getting back to the wrong nature' in W. Cronon (Ed) *Uncommon Ground: Rethinking the Human Place in Nature*. New York: W.W. Norton and Co, pp. 69–90.

Crowley, S. L., Hinchliffe, S. and McDonald, R. A. (2017) 'Nonhuman citizens on trial: The ecological politics of a beaver reintroduction', *Environment and Planning A*, 49(8), pp. 1846–1866.

Deary, H. and Warren, C. R. (2018) 'Trajectories of rewilding: A taxonomy of wildland management', *Journal of Environmental Planning and Management*, 62(3), pp. 466–491.

DeMello, M. (2012) *Animals and Society: An Introduction to Human-animal Studies*. New York: Columbia University Press.

DeSilvey, C. and Bartolini, N. (2018) 'Where horses run free? Autonomy, temporality and rewilding in the Côa Valley, Portugal', *Transactions of the Institute of British Geographers*, 44, pp. 94–109.

Donlan, J. (2005) 'Re-wilding North America', *Nature*, 436(18), pp. 913–914.

EIB (2018) *European Investment Bank Environmental and Social Standards*. Luxembourg. Available at: https://www.eib.org/attachments/strategies/environmental_and_social_practices_handbook_en.pdf

Fiege, M. (1999) *Irrigated Eden: The Making of an Agricultural Landscape in the American West*. Seattle: University of Washington Press.

Fisher, M. (2004) 'Self-willed land: can nature ever be free?', *Ecos*, 25(1), pp. 6–11.

Fisher, M. and Parfitt, A. (2016) 'The challenge of wild nature conserving itself', *Ecos*, 37, pp. 27–34.

Foreman, D. (1992) 'Around the campfire', *Wild Earth*, 2(4), pp. 1–2.

Frankel, O. H. and Soule, M. E. (1981) *Conservation and Evolution*. Cambridge: Cambridge University Press.

Gammon, A. R. (2018) 'The many meanings of rewilding: An introduction and the case for a broad conceptualisation', *Environmental Values*, 27(4), pp. 331–350.

Garrard, G. (2020) 'Brexit ecocriticism', *Green Letters*, 24(2), pp. 110–124.

Gillson, L., Ladle, R. J. and Araújo, M. B. (2011) 'Baselines, patterns and process' in Ladle, R.J. and Whittake, R. J. (Eds) *Conservation Biogeography*. Chichester: Blackwell.

Green, B. (1996) *Countryside Conservation*. 3rd ed. London: Chapman and Hall.

Hall, M. (2014) 'Extracting culture or injecting nature? Rewilding in transatlantic perspective' in M. Drenthen and J. Keulartz (Eds) *Old World and New World Perspectives in Environmental Philosophy*. London: Springer International Publishing, pp. 17–35.
Haraway, D. (2003) *The Companion Species Manifesto: Dogs, People and Significant Otherness*. Chicago: Prickly Paradigm Press.
Harrison, M. (2016) 'Rewilding National Parks: Moor than meets the eye', *Ecos*, 37(1), pp. 43–47.
Harthoorn, A. M. (1962) 'Translocation as a means of preserving wild animals', *Oryx*, 6(4), pp. 215–227.
Head, L. (2015) 'The Anthropoceneans', *Geographical Research*, 53(3), pp. 313–320.
Helmreich, S. (2005) 'How scientists think; about "natives", for example. A problem of taxonomy among biologists of alien species in Hawaii', *Journal of the Royal Anthropological Institute*, 11, pp. 107–128.
Hintz, J. (2007) 'Some political problems for rewilding nature', *Ethics, Place and Environment*, 10(2), pp. 177–216.
Höchtl, F., Lehringer, S. and Konold, W. (2005) '"Wilderness": What it means when it becomes a reality - A case study from the southwestern Alps', *Landscape and Urban Planning*, 70(1–2), pp. 85–95.
Hodder, K. H. et al. (2005) *Large Herbivores in the Wildwood and in Modern Naturalistic Grazing Systems*. Peterborough. Available at: publications.naturalengland.org.uk/file/93025
Hodder, K. H. et al. (2009) 'Can the pre-Neolithic provide suitable models for rewilding the landscape in Britain?', *British Wildlife*, 20(5), pp. 4–15. Available at: http://eprints.bournemouth.ac.uk/10323/4/licence.txt
Hoskins, W. G. (1955) *The Making of the English Landscape*. London: Penguin.
Ingold, T. (2000) *The Perception of the Environment*. London: Routledge.
IUCN (2013) *Guidelines for Reintroductions and Other Conservation Translocations*. Gland: IUCN. Available at: https://portals.iucn.org/library/efiles/documents/2013-009.pdf
IUCN (2019) *About the IUCN*. Gland: IUCN. Available at: https://www.iucn.org/about (Accessed: 29 January 2019).
IUCN CEM (2021) *Rewilding Principles*. Gland: IUCN. Available at: https://iucn.org/sites/default/files/2022-10/principles_of_rewilding_cem_rtg.pdf
Jepson, P. (2016) 'A rewilding agenda for Europe: Creating a network of experimental reserves', *Ecography*, 39, pp. 117–124.
JNCC (1996) *Review of Information, Policy and Legislation on Species Translocation*. Peterborough: JNCC. Available at: https://citeseerx.ist.psu.edu/document?repid=rep1&type=pdf&doi=1cb82ca11387e92ea76f5ca64d24e56c786d2669
JNCC (2015) *JNCC - who we are*. Available at: http://jncc.defra.gov.uk/default.aspx?page=5287 (Accessed: 29 January 2019).
John Muir Trust (2020) *Glenlude*. Available at: https://www.johnmuirtrust.org/trust-land/glenlude (Accessed: 21 June 2025).
Jørgensen, D. (2015) 'Rethinking rewilding', *Geoforum*, 65, pp. 482–488.
Keulartz, J. (2009) 'Boundary Work in Ecological Restoration', *Environmental Philosophy*, 6(1), pp. 35–56.
Knepp Wildland (2020a) *Background*. Available at: https://knepp.co.uk/background (Accessed: 21 June 2025).
Knepp Wildland (2020b) *Restoration of Repton Park*. Available at: https://knepp.co.uk/knepp-estate/history/restoration-of-repton-park/ (Accessed: 21 June 2025).
Latour, B. (1993) *We Have Never Been Modern*. London: Harvester Wheatsheaf.
Light, A. (1995) 'Urban wilderness' in D. Rothenberg (Ed) *Wild Ideas*. Minneapolis: University of Minnesota Press, pp. 195–211.

Linnell, J. D. C. et al. (2015) 'Framing the relationship between people and nature in the context of European conservation', *Conservation Biology*, 29(4), pp. 978–985.

Lorimer, J. and Driessen, C. (2014) 'Wild experiments at the Oostvaardersplassen: Rethinking environmentalism in the Anthropocene', *Transactions of the Institute of British Geographers*, 39(2), pp. 169–181.

Lorimer, J. et al. (2015) 'Rewilding: Science, practice, and politics', *Annual Review of Environment and Resources*, 40, pp. 39–62.

Lovelock, J. (1979) *Gaia*. Oxford: Oxford University Press.

Maffi, L. and Woodley, E. (2010) *Biocultural Diversity Conservation*. Abingdon: Taylor & Francis.

Martin, P. (1992) 'The Last Entire Earth', *Wild Earth*, 2(4), pp. 29–32.

Matless, D. (1998) *Landscape and Englishness*. London: Reaktion Books Limited.

Matless, D. (2003) 'Landscape: Introduction' in K. Anderson et al. (Eds) *Handbook of Cultural Geography*. London: SAGE Publications, pp. 227–232.

McKibben, B. (1990) *The End of Nature*. London: Penguin.

Merckx, T. and Pereira, H. M. (2015) 'Reshaping agri-environmental subsidies: From marginal farming to large-scale rewilding', *Basic and Applied Ecology*, 16(2), pp. 95–103.

Mitchell, D. (1994) 'Landscape and surplus value: the making of the ordinary in Brentwood, CA', *Environment and Planning: Society and Space*, (12), pp. 7–30.

Mitchell, D. (1996) *Lie of The Land: Migrant Workers and the California Landscape*. Minnesota: University of Minnesota Press.

Monbiot, G. (2013) *Feral*. London: Penguin.

Morrow, J. (2011) 'Romanticism and political thought in the early nineteenth century' in G. Stedman Jones and G. Claeys (Eds) *The Cambridge History of Nineteenth-Century Political Thought*. Cambridge: Cambridge University Press, pp. 39–76.

Navarro, L.M. and Pereira, H. M. (2012) 'Rewilding Abandoned Landscapes in Europe', *Ecosystems*, 15(6), pp. 900–912.

Newton, M. (2019) *From Summit to Sea: Natural solutions to decarbonising landscapes*. Available at: https://www.endangeredlandscapes.org/news/from-summit-to-sea-natural-solutions-to-decarbonising-landscapes/ (Accessed: 21 June 2025).

Noss, R. (1994) 'Wilderness-now more than ever', *Wild Earth*, 4(4), pp. 187–194.

Nye, D. (2000) 'Technologies of landscape' in D. Nye (Ed) *Technologies of Landscape: From Reaping to Recycling*. Massachusetts: University of Massachusetts Press, pp. 3–17.

OED (2020a) *Rewild, Oxford English Dictionary Online*. Oxford: Oxford University Press. Available at: https://www.oed.com/search/dictionary/?scope=Entries&q=rewild

OED (2020b) *Translocation, Oxford English Dictionary Online*. Oxford: Oxford University Press. Available at: https://www.oed.com/search/dictionary/?scope=Entries&q=translocation

Olwig, K. R. (2002) *Landscape, Nature, and the Body Politic: From Britain's Renaissance to America's New World*. Madison: University of Wisconsin Press.

Olwig, K. R. (2016) 'Virtual enclosure, ecosystem services, landscape's character and the "rewilding" of the commons: the "Lake District" case', *Landscape Research*, 41(2), pp. 253–264.

Pettorelli, N. et al. (2018) 'Making rewilding fit for policy', *Journal of Applied Ecology*, 55, pp. 1114–1125.

Piesing, M. (2016) 'George Monbiot on "rewilding" the countryside and the end of sheep', *i*, October. Available at: https://inews.co.uk/inews-lifestyle/george-monbiot-rewilding-countryside-end-sheep-23999

Prior, J. and Brady, E. (2017) 'Environmental aesthetics and rewilding', *Environmental Values*, 26(1), pp. 31–51.

Prior, J. and Ward, K. J. (2016) 'Rethinking rewilding: A response to Jørgensen', *Geoforum*, 69, pp. 132–135.

Sandom, C. J., Hughes, J. and Macdonald, D. W. (2013) 'Rooting for rewilding: Quantifying wild boar's sus scrofa rooting rate in the Scottish Highlands', *Restoration Ecology*, 21(3), pp. 329–335.

Sandom, C. J. and Macdonald, D. W. (2015) 'What next? Rewilding as a radical future for the British countryside' in D. W. Macdonald and R. E. Feber (Eds) *Wildlife Conservation on Farmland Volume 1: Managing for Nature in Lowland Farms*. Oxford: Oxford University Press, pp. 291–316.

Sandom, C. J. and Wynne-Jones, S. (2019) 'Rewilding a country: Britain as a study case' in N. Pettorelli, S. Durant and J. du Toit (Eds) *Rewilding*. Cambridge: Cambridge University Press, pp. 222–247.

Sandom, C. J. et al. (2013) 'Rewilding', in D. W. Macdonald and K. J. Willis (Eds) *Key Topics in Conservation Biology 2*. Chichester: John Wiley & Sons, pp. 430–451.

Sandom, C. J. et al. (2018) 'Rewilding in the English uplands: Policy and practice', *Journal of Applied Ecology*, 56, pp. 266–273.

Saunders, G. (2016) 'Family ties Human views of nature -time to see goodness before wildness?', *Ecos*, 37(3–4), pp. 49–54.

Schein, R. (1997) 'The place of landscape: A conceptual framework for interpreting an American scene', *Annals of the Association of American Geographers*, 87(4), pp. 660–680.

Seddon, P. J. et al. (2014) 'Reversing defaunation: Restoring species in a changing world', *Science*, 345(6195), pp. 406–412.

Sheail, J. (2007) '"One and the same historic landscape": A physical/cultural perspective', *International Journal of Heritage Studies*, 13(4–5), pp. 321–334.

Short, L. (2002) 'Wind power and English landscape identity' in M. Pasqualetti (Ed) *Wind Power in View: Energy Landscapes in a Crowded World*. London: Academic Press, pp. 43–58.

Shucksmith, M. (2018) 'Re-imagining the rural: From rural idyll to Good Countryside', *Journal of Rural Studies*, 59, pp. 163–172.

Soule, M. and Noss, R. (1998) 'Rewilding and biodiversity: Complementary goals for continental conservation', *Wild Earth*, 8(3), pp. 18–28.

Stoppard, T. (1993) *Arcadia*. London: Faber and Faber.

Svenning, J.C. et al. (2016) 'Science for a wilder Anthropocene: Synthesis and future directions for trophic rewilding research', *Proceedings of the National Academy of Sciences of the United States of America*, 113(4), pp. 898–906.

Tanasescu, M. (2017) 'Field notes on the meaning of rewilding', *Ethics, Policy and Environment*. Routledge, 20(3), pp. 333–349.

Taylor, P. (2004) 'Towards a wildland strategy', *Ecos*, 25(3/4), pp. 1–3.

Taylor, P. (2005) *Beyond Conservation: A Wildland Strategy*. London: Earthscan Publications Limited.

Thompson, I. (2010) *The English Lakes: A History*. London: Bloomsbury.

Toledo, D., Agudelo, M. S. and Bentley, A. L. (2011) 'The shifting of ecological restoration benchmarks and their social impacts: Digging deeper into Pleistocene re-wilding', *Restoration Ecology*, 19(5), pp. 564–568.

Tree, I. (2018) *Wilding*. London: Picador.

Tree, I. and Burrell, C. (2023) *The Book of Wilding: A Practical Guide to Rewilding, Big and Small*. London: Bloomsbury Publishing.

Tsing, A. (2012) 'Unruly edges: Mushrooms as companion species', *Environmental Humanities*, (1), pp. 141–154.

van der Zanden, E. H. et al. (2017) 'Trade-offs of European agricultural abandonment', *Land Use Policy*, 62, pp. 290–301.

Vera, F. (2000) *Grazing Ecology and Forest History*. Wallingford: CABI Publishing.

White, R. (1995) '"Are you an environmentalist or do you work for a living?": Work and nature' in W. Cronon (Ed) *Uncommon Ground: Rethinking the Human Place in Nature*. New York: W.W. Norton and Co, pp. 171–185.

Wilderness Act (1964) *The Wilderness Act*. USA.
Williams, R. (1975) *The Country and The City*. London: Chatto and Windus.
Wittgenstein, L. (1968) *Philosophical Investigations*. London: Blackwell.
Wright, J. P., Jones, C. G. and Flecker, A. S. (2002) 'An ecosystem engineer, the beaver, increases species richness at the landscape scale', *Oecologia*, 132, pp. 96–101.
Wylie, J. (2007) *Landscape*. Oxon: Routledge.
Wynne-Jones, S., Strouts, G. and Holmes, G. (2018) 'Abandoning or reimagining a cultural heartland? Understanding and responding to rewilding conflicts in Wales - The case of the Cambrian Wildwood', *Environmental Values*, 27(4), pp. 377–403.
Zeller, T. (2007) *Driving Germany: The Landscape of the German Autobahn, 1930–1970*. New York: Berghahn Books.

2 'Rewilding' the Avalon Marshes

The Avalon Marshes landscape

Rising above the marshes, Glastonbury Tor, the mythical Isle of Avalon, is visible on the horizon long before one reaches it. Nor does it appear to get any closer, lost as the traveller is in the marshes and a maze of droves[1] which resemble nothing so much as an Escher drawing. The droves lead the traveller ever on but somehow never closer to the Tor, always perpendicular to it, so that one has to travel miles out of one's way before reaching a T-junction or crossroads, each one indistinguishable from the last. The droves, though arrow straight, are in no way flat despite, or rather because of, their crossing the flatlands of the marshes. They buck and hummock skywards as the peat beneath them shifts and sinks, their edges seeping gradually from tar to grass and water. At the same time the grass makes incursions in the other direction, encroaching on the tarmac as part of the slow, diffusive dance which is seeing wildlife reclaiming what was once human territory.

This chapter discusses the 'rewilding' of the Avalon Marshes – low-lying wetlands near Glastonbury, Somerset in the southwest of England which form part of the larger Somerset Levels and Moors. The chapter introduces the landscapes of the Avalon Marshes and discusses themes which emerged during interviews with rewilding stakeholders and practitioners. Key among those stakeholders are the Avalon Marshes partners: the Hawk and Owl Trust (HOT), the Somerset Wildlife Trust (SWT), the Royal Society for the Protection of Birds (RSPB) (all wildlife conservation charities) and Natural England (a public body to advise the government and protect the environment). Together, these organisations own the five nature reserves which collectively constitute the Avalon Marshes: the Catcott Complex (owned by the SWT), Ham Wall (owned by the RSPB), Shapwick Heath (owned by Natural England), Shapwick Moor (owned by the HOT) and Westhay Moor (owned by the SWT). It is the land and water in these reserves (some 1500 hectares) which forms the focus of this chapter.

DOI: 10.4324/9781003456490-3

The naming of rewilding in the Avalon Marshes

In discussing the Avalon Marshes with respect to rewilding it should be made explicit that the Avalon Marshes do not self-identify as rewilding. Conservation efforts in the area precede the trend for rewilding as a practice and a term and, as the term has come into vogue, conservation practitioners in the Avalon Marshes have consciously avoided it. This was made very clear by Niall, a conservationist, who said that the Avalon Marshes reserves were 'around a long time before rewilding was made popular as a concept' and that 'it would have been insincere to try and cash in on' the term. In particular Niall argued that rewilding is a new term for activities, and even concepts, which pre-dated it: 'the concepts behind it [rewilding] about restoring and creating habitat and big landscapes for wildlife, that was happening, the concept was there already, so I think rewilding is a new term for something that's already been happening'. While some scholars would argue that rewilding is fundamentally different from conventional conservation, these comments echo other scholars who suggest that rewilding is something of a rebranding of what might previously have been called ecological restoration (e.g., Murray, 2017).

Importantly, however, many stakeholders and practitioners highlighted aspects of practice in the Avalon Marshes which precluded the conservation effort there from being considered rewilding – in particular the level of human intervention. As discussed in Chapter 1, human intervention is complex in this regard: on one hand, understandings of 'active rewilding'[2] (Carver, 2016; Sandom et al., 2018) permit and even require significant human intercession to address the impact of previous human activity and/or to catalyse natural processes which would otherwise be slow and gradual, while other interpretations insist that rewilding should not involve human intervention. As with much in relation to rewilding, however, these interpretations fall along a spectrum from no human intervention being acceptable (Höchtl, Lehringer and Konold, 2005), to human intervention which is reduced or at the very least reducing being acceptable (Navarro and Pereira, 2012; Lorimer et al., 2015; Pettorelli et al., 2018). Octavia, a conservationist, gave an excellent illustration of this in relation to the Avalon Marshes, although it is notable that they used the term wilding rather than rewilding:

> the wilding bit of it [the Avalon Marshes] is that most of the large-scale habitats in the Avalon Marshes reserves are old peat diggings that have been restored; the initial restoration was directed towards conservation outcomes and they have been restored to a more or a less defined version of wildlife habitats.

This comment draws attention to interpretations of rewilding which include the restoration of land which was used for, and potentially significantly ecologically degraded by, human activity, and returning it to nature

(Nogués-Bravo et al., 2016; DeSilvey and Bartolini, 2018; Gammon, 2018; IUCN, 2021). This was certainly the case in the Avalon Marshes where, as Octavia notes, land which had previously been used for peat extraction was given over to conservation.

Another important issue raised by this comment is the fact that sites of former peat extraction have been 'restored to a more or a less defined version of wildlife habitats'. This hints at the specific outcomes which the restoration was aiming for and which rewilding commonly eschews. It is true that the RSPB, who were involved in some of the earliest stages of the restoration of the Avalon Marshes, were specifically attempting to create habitat favourable to bitterns (*Botaurus stellaris*). From this perspective, restoration of the Avalon Marshes could be seen as conventional, target-driven conservation rather than rewilding, which is classically seen as *not* being goal orientated (Gammon, 2018). The habitat creation has, however, been attractive to many species beyond the bittern and has therefore prompted 'auto-rewilding' (Clancy and Ward, 2020); auto-rewilding is the natural recolonisation of an area via self-reintroduction, based on 'the evolutionary responses of species' (Gillson, Ladle and Araújo, 2011). Auto-rewilding relies on, and manifests, the increase in other-than-human agency, which rewilding aims to promote (Arts, Fischer and van der Wal, 2016). The recolonisation of the Avalon Marshes by birds was mentioned by several participants who highlighted their agency in returning of 'their own accord', finding 'their own way here' and 'returning by themselves'. This auto-rewilding demonstrates the other-than-human agency at work in the Avalon Marshes and contributes to the classification of the Avalon Marshes as rewilding.

Despite not self-identifying as rewilding, the Avalon Marshes *are* described as rewilding by others (e.g., Moss, 2016; Taylor, 2017; Macdonald, 2019) including Robert, a rewilding author/journalist, who said they would 'be interested to see if the RSPB think Ham Wall is an example of rewilding. *I would say it is but then I take rewilding in a very wide sense*'. What is particularly interesting about this comment is, first, that Robert (correctly) identifies that the Avalon Marshes do not self-identify as rewilding, and second, that they acknowledges that their own view of rewilding interprets the term in a 'wide sense' – an implicit acknowledgement that the RSPB (which manages Ham Wall) and the other conservation organisations involved in the Avalon Marshes may have a narrower interpretation than this. Nonetheless, as Niall put it, 'the Avalon Marshes isn't a rewilding scheme but essentially that's what's happened here – wildlife is thriving because it's a big landscape, all joined up and connected, it's doing really well'.

The practice of rewilding in the Avalon Marshes

In addition to justifying the classification of the Avalon Marshes as a case of rewilding, the discussion in the previous section has given hints of the rewilding activity taking place there: ecological restoration on a landscape scale, restoring land from human use (conventional agriculture and industrial peat extraction) to wild(er) ecosystems which are, at times, combined with extensive farming. The key human and other-than-human activities involved in this process are habitat creation via the establishment of reedbeds, low-intensity grazing with cattle, goats and ponies, changes in drainage and water level management, changes in soil management and species recolonisation.

Humans have been intervening in the landscapes of the Avalon Marshes for millennia and the landscape has undergone significant changes, and changes of use, over the years. One of the most significant of these was the development and subsequent decline of the peat industry. Peat digging by hand had traditionally been carried out in and around the Avalon Marshes to obtain peat for burning. In the 20th century there was a shift from the use of peat for burning to its use in horticulture and an associated industrialisation of peat extraction. Industrial peat extraction is essentially a form of surface mining, removing layers of peat leaving large, shallow pits in the landscape. The resulting landscape is, self-evidently, intensely anthropogenic, having been significantly shaped by human activity. It was this landscape which was left behind when large-scale peat extraction began to decline in the 1990s following an embargo on new permissions to extract peat or extensions to existing extraction permissions. This effectively constituted productive land abandonment (Jørgensen, 2015), with one major peat extraction company ceasing resource exploitation and relinquishing ownership of much of the land. That land was handed to English Nature (now Natural England) which in turn transferred land to the SWT and the RSPB. An early part of rewilding in the Avalon Marshes was to transform the post-industrial landscape of peat extraction to wildlife habitat.

In the initial stages, this transformation involved a period of highly interventionist activity to modify the land which had been acquired. In the case of Ham Wall, this involved hand-planting large numbers of reeds in worked out peat voids to create reedbeds as wildlife (particularly *bird*life) habitat. In the case of Shapwick Moor, it involves the (attempted and ongoing) reduction in soil nutrients and the rewetting of peat to convert intensive farmland into wildflower meadows. While in many places the first phase of intensive activity has ceased, or at least decreased, every reserve of the Avalon Marshes continues to be managed to some degree, in a slow, ongoing and complex negotiation between humans, land and water.

This is particularly true in relation to water with rewilding changing centuries- old human–water relationships. Previous human interactions with water in the marshes primarily focussed on attempts to increase drainage, improving the land for agriculture. Rewilding is reversing that relationship, attempting to reduce drainage, raise the water table and allow or encourage pools to form. The painstaking process was described by Alan, a conservationist: 'to bring this back to a wet site is quite complicated, but we're slowly doing it now, we're damming ditches, cutting land drains, digging scrapes, but it's a very expensive and slow process, it's still going to take many years'. What becomes apparent from a study of the Avalon Marshes, and comments such as this, is that its landscapes, perhaps most especially the reedbeds, are a *relational achievement* (Whatmore, 1999) in the creation of place between the creative agencies of the peat soil, water, reeds, birds, humans and technology.[3] The reeds have what Cloke and Jones (2001) would call a unique creative ability to form marshes in negotiation with water, which humans and birds exploit in turn – the humans exploiting the ability of the reeds and the birds exploiting the resultant marshland.

In creation of place, such as the rewilding of the Avalon Marshes, we can also consider the relational achievements that occur when the innate characteristics of one agent result in co-dependent shaping of and by other agents (Cloke and Jones, 2001). The peat soil of the Avalon Marshes provides an excellent example of this. The affordances of peat as fuel and compost have shaped the lives of the people in the Avalon Marshes – providing work digging the peat, shaping digging practises and (increasingly mechanised) digging technology. Digging has, in turn, shaped the peat – drying it, causing it to contract and sink and leaving peat-shaped voids in the landscape where the peat used to be. In turn, the embargo on the issuing of further permission to extract peat has left a void in the landscape where the peat industry used to be, so that there is only a remnant peat industry and, in many places, peat industry-shaped holes in the landscape. This void has, in turn, been filled by wildlife, as water, reeds, birds and people have worked together to co-create and shape the marshes, each exerting their own agency. This agency was seen as particularly profound in the case of water with Octavia, a conservationist, saying that in the case of wetland restoration 'water does a lot of the work for you'. Here, humans are making water a proxy for them, in a similar way as they do with other-than-human animals as will be discussed in Chapter 5. Robert, a rewilding author/journalist, made a related point saying that 'wetlands are very quick to do, just add water and within five years you have a nature reserve'. This comment highlights the agency of water as a catalyst and also its alchemy – its ability to transform a degraded landscape into wetland habitat.

The perception and negotiation of rewilding in the Avalon Marshes

As well as understanding how rewilding is practiced in the Avalon Marshes, it is important to understand how it is perceived. It is, however, very difficult to disentangle perceptions of rewilding from attitudes towards a plethora of other environmental concerns in the area including the embargo on new permissions to extract peat, the original establishment of the nature reserves which now constitute the Avalon Marshes, the designation of land as sites of special scientific interest, water management (particularly changes to dredging protocols) and government payments to farmers in relation to environmentally sensitive farming. Notwithstanding this, participants did identify a general change in attitudes towards these concerns, and a subsequent shift in the way in which they are negotiated. For example, Wiliam, an archaeologist, compared current negotiations over rewilding to controversies which have occurred in relation to conservation in the past saying that there is 'a lot less controversy nowadays because there's more realisation that things are going to change, and sustainability is a much more important issue for everyone so I think there's generally more co-operation'.

From a stakeholder point of view, this reduction in controversy stemmed not so much from an acceptance of conservation demands as a shift in the centre of gravity in negotiations. They considered that it reflected what is perhaps best described as a lack of balance – with the perception being that land management, and regulations around land management, had moved (too far) in favour of conservation (including rewilding) at the expense of other, productive, land uses. Balance is a recurring theme in conservation debates, especially when human and other-than-human interests are opposed. Cassidy (2019) identified it in the debate concerning badgers and bovine tuberculosis, for example, with farmers seeing their role as stewards in maintaining 'natural balance' as being hampered by 'overly strict legislation'. In the Avalon Marshes, stakeholders perceived the increase in consideration being given to conservation as an unbalancing of the power between it and other rural land uses saying 'I don't think they've got the balance right, not at all', 'the balance is going the other way', and 'we realise the balance is tipping the wrong way'. These last two comments are particularly interesting as they reveal viewpoints which would appear to privilege farming and fishing (these stakeholders were involved in agriculture and angling, respectively) over nature, conservation and rewilding, and suggest that the 'balance' used to be in favour of farming and fishing. David, a peat/compost producer, saw the change in balance as making conservationists and conservation organisations 'much harder to deal with' and suggested that increased consideration, or preference, was being given to conservation over other concerns, saying 'it's just conservation, conservation, conservation all the time, even the land we've got, we can't even make a fishing lake out of it for fishing, it's got to be for conservation'. This point was echoed by Larry, a landowner/manager who said, 'I think conservation has definitely got more importance than it used to and it's definitely a much bigger area than it always was, so that just changes the impact a little bit'. This perception is reflected by scholars who trace it back to overproduction in European agriculture in the 1980s which afforded new 'possibilities for conservation' (McEachern, 1992). Prior to that, even National Parks had to proceed 'sensitively' in negotiations with agriculture due to their 'limited powers' (McEachern, 1992). This contrasted with the significant power and influence of farmers and their lobby groups (the National Farmers Union and the Country Land and Business Association) which had previously prevailed, even in the face of environmental concerns (McEachern, 1992). Appositely for this discussion, the example McEachern gives is of the Somerset Levels where 'particularly influential farmers were unwilling to relinquish power and control' (1992, p. 161). More recently, other scholars have identified that changes in conservation policy have reduced landowner 'ability to leverage' in negotiations with conservation (Lennox et al., 2012). This supports the views of the stakeholders quoted above that conservation is becoming more successful (or at least more powerful) in negotiating its boundaries with other land uses and has increased its influence in these negotiations in relation to other parties. Thus, while from the practitioner perspective, negotiations regarding conservation, and by extension rewilding, have got easier (because public awareness regarding sustainable land management and resource use has increased), from the stakeholder point of view, negotiations have got harder for the same reason – that is because greater weight is given to conservation concerns.

In relation to rewilding specifically, there is a critique that it may disadvantage specialist species which have specific requirements in terms of habitat, diet or nesting. This concern was present among stakeholders who noted the changes in the species present in the Avalon Marshes. This was accompanied by the perception that rewilding had not necessarily been good for all species, with stakeholders suggesting that certain species (including barn owls and ground nesting birds) had disappeared from the landscape, leading them to view rewilding somewhat negatively. Larry, a landowner/manager, suggested that the rewilding landscape, particularly the wetter landscape, did not suit species which were there previously:

> when the water levels are kept higher in the spring, the land lies wetter, your ground nesting birds won't nest, birds don't nest in water: that land, it was never wet, or not as wet for as long as it is now, it tended to be drier, so I think that has an impact, a big impact on certain bird life.

Larry went on to say that vulnerable species in particular had suffered, describing nature as cruel: 'nature is a cruel thing really, survival of the fittest, and when certain animals move in and they have a free rein your vulnerable ones are going to go, and unfortunately I think that's what's happened'. The language of this comment reveals a negative view of rewilding in which vulnerable species are lost to cruel nature without the managing hand of man to keep it in check. This view was supported by David, a peat/compost producer, who talked about species being lost from the landscape and an overall decline in biodiversity, despite an increase in bioabundance:

> you don't see many barn owls about because when it's all flooded and it's all wet there's no mice for the barn owls to feed on … [before the rewilding of the Avalon Marshes] there was more of a diversity of wildlife than there is now. There's a lot more wildlife now but not the diversity.

These comments reflect criticism of rewilding by scholars who argue that rewilding will result in biodiversity loss in the medium and long term with rare species particularly at risk (Navarro and Pereira, 2012; Lasanta, Nadal-Romero and Arnáez, 2015; Sandom et al., 2018).

Rewilding's negotiation with peat extraction in the Avalon Marshes

The peat soil of the Avalon Marshes has created a subterranean landscape with a complex past and present. Peat has been dug in the area for centuries, originally as fuel for burning but, by the middle of the 20th century, this had been almost entirely replaced by the extraction of peat for horticulture. Not only was the peat being used for a different purpose, its extraction was very different, being dug mechanically rather than by hand as it had been traditionally. This wholesale extraction of peat had two significant environmental impacts, one local and the other global. Locally, it created peat voids, large areas of land from which significant amounts of peat had been removed in a process resembling opencast mining, the result being crater-like holes in the landscape. The other impact, which had global effects, was the release of carbon dioxide (a potent greenhouse gas) when the peat was drained and dried. Knowledge of this latter effect contributed to a decision by the Ministry of Housing, Communities and Local Government (MHCLG) which saw planning permission for peat extraction halted for new sites and not extended for existing sites (MHCLG, 2012, 2019), a policy which is the culmination of a strategy to 'reduce peat use to zero by 2030' (DEFRA, 2011). It should be noted that the embargo on the granting of planning permission for peat extraction applies only to new sites or extensions to exiting sites; sites which already have planning permission are currently permitted to continue extracting peat in accordance with their existing permissions, with many peat producers in Somerset falling into this category. It should also be noted that the embargo is a national decision and not one which is local to the Avalon Marshes or related specifically to rewilding in the area.

The response of Somerset County Council (SCC) with regard to local planning and regulations has, naturally, been in line with national government 'Restoration and Aftercare of Minerals Sites' guidelines (SCC, 2013). These guidelines state that the restoration of land where mineral extraction has taken place 'should use every opportunity to enhance the environmental value of sites to contribute to the biodiversity of the County' (SCC, 2013). Accordingly, SCC have developed the 'Somerset Minerals Plan' which, *inter alia*, 'seeks to ensure any scheme for restoration of a former peat workings will focus on promoting nature conservation and biodiversity' (SCC, 2015). This approach, coupled with the embargo on new extraction of peat has contributed to the complexity of the relationship between rewilding and peat extraction. While the rewilding at the Avalon Marshes is not responsible for the embargo, it is benefiting from the phasing out of the industry. This has led to some stakeholders conflating the rewilding of the area with the ending of new peat extraction licenses, as perhaps did the fact that the Avalon Marshes are a very visible representation of conservation locally.

One major point for rewilding to negotiate in this context is the economic situation – peat production provides a significant source of income to the area which rewilding currently appears unable to replace. Several participants spoke of the financial benefits of peat production and the livelihoods, and indeed way of life, that it supports. Caroline, a conservationist, described how the Avalon Marshes 'was a place where lots of money was made once ... some will harken back to those halcyon days'. This view was supported by other participants who described what a big employer the peat industry had been and continued to be. Indeed, Dylan, another conservationist, acknowledged how hard it would be to replace those jobs and livelihoods: 'all these local peat workers, they are local people, this is their livelihood, what are they going to do instead?'. Victor, a peat/compost producer, was particularly articulate and vociferous on this point stating that, if peat extraction ceased, 'nothing you can do afterwards will ever generate the same level of economic activity out there in the wider community'. Victor went on to stress that conservation is a poor economic alternative to consumptive land use generally but particularly as compared to peat extraction:

> the issue with restoration to conservation [post peat extraction] is you're taking land that had an economic value to the community as a whole, not just to the owner, to the community as a whole, in so much as the revenue that it generates, the multipliers out there in the wider economy are much greater, and then you're going to a conservation after use, a restoration, which has a value to society that can be monetised, but it's not a monetary value, and what you've seen in this area is you've taken a big chunk of land out of a productive land use, you've taken a big chunk of land out of something that generated revenues for the local community and put it into something that has very limited revenue generation.

This argument is very similar to that raised by scholars who question if and how the ecotourism associated with conservation might be able to replace the employment and incomes generated by agriculture.

The fact that the Avalon Marshes have evolved from the decline of the peat industry is an inescapable fact and one which was often highlighted by participants. In the case of stakeholders, particularly, peat producers, there appeared to be a suggestion that the peat industry deserved credit for this or, perhaps, that because the Avalon Marshes exist on what was peat extraction land, the Avalon Marshes were beholden to the peat industry, and even that their continued existence depended on the sufferance of the peat industry. Ian, a peat/compost producer, illustrated this saying:

> they say, 'oh the Avalon Marshes,' but they've never once said what brought 'em about, and that's what they don't realise, if that land had been never dug for peat, they wouldn't be there, and I'm sure if we'd said, 'oh we're going to take 'em away,' they wouldn't like it.

Ian places emphasis on the peat industry being a precursor to the Avalon Marshes as if it were a necessity for their creation. Ian went on to assign some of the credit for the recolonisation of the area by wildlife to the peat industry as a defence against those whom, they say, claim peat extraction is 'ruining the area':

> when people say, 'oh you're ruining the area by peat extraction,' I think they're totally wrong. They've got to make a decision: do they want wildlife here or don't they? And if they don't want wildlife well no peat extraction but if they want wildlife, which there is, that's how the wildlife come, by the peat extraction … if we hadn't extracted the peat, the wildlife wouldn't be there. So, I think they're totally wrong when they say, 'when you dig the peat, you ruin the area'.

George, another peat/compost producer, made a related, though less contentious, point when he credited the peat industry with creating a 'beautiful, beautiful area' part of which he attributed to the creation and modification of the drainage waterways:

> this river wouldn't be here if it wasn't for the original peat diggers that did burning blocks, you would just have the main river that flows through Glastonbury and Street, the river Brue, and then that disappearing down to Highbridge and into the sea. But now we have this massive water way, water area, on the back of the peat diggers.

The river this participant is referring to, and considers so beautiful, is the rather unfortunately named South Drain, and is a canal rather than a river, since, as they identify, it was created by human processes. A similar aesthetic appeal has been attributed to the Norfolk Broads which, like the Avalon Marshes, have a history of peat digging (Matless, 2014). An evocative article from a Norfolk newspaper in 1953 sums up the juxtaposition of the 'artificiality' and the beauty of these landscapes, 'so seemly are they in our landscape that it is with something like a shock that one learns how much the great majority owe to man's handiwork; the beauty is *artificial* rather than *natural*' (Matless, 2014, p. 47 emphasis added). This quote raises an interesting question with regard to rewilding and the way landscapes and nature are valued and considered authentic seeming, as it does, to suggest that the artificiality diminishes the beauty in some way. Rewilding, with its 'creation' of nature has to negotiate a balance between the 'artificial' and the 'natural' and *re*negotiate the way that nature is valued since, if nature is created via human processes rather than developing via 'natural' processes it may not be considered authentic and may therefore be valued differently (Katz, 1998).

Irrespective of its future, the peat industry is, for the time being, continuing in the Avalon Marshes area and, in some respects, rewilding and peat extraction can coexist harmoniously. Dylan, a conservationist, emphasised the way that rewilding sites exist adjacent to peat extraction land:

> where we're sat right now [Westhay Moor], not 100 yards over to the left is a peat works where they are still digging peat and, as you cycled round and along the road, there was an area that's just coming to the end of peat digging. So, straight away you've got that within the landscape that we are in. So, you have peat digging still going on, as it will do for the next twenty years or so.

Stakeholders expressed a desire for this coexistence between humans and nature to continue, and for *peat extraction* to be allowed to continue, particularly with regard to providing livelihoods which they saw as otherwise excluded by rewilding, or conservation generally with David, a peat/compost producer saying:

> I think that if more planning permission was granted for the excavation of peat on certain land where it wasn't high conservation value, then at the end of the day, when the peat land was dug out, it could then be turned over to Natural England and so it could then again be turned back for wildlife purposes. I don't think there'd be a problem on that, there certainly wouldn't be a problem on that, we'd be willing to do that. We were offered twelve acres of land right next to our workings six months ago. We had no hope of getting planning permission on it. It used to be a digging zone, it's right in the middle of the digging zone. We would willingly give that land back when it was dug out to Natural England.

Allowing such continued coexistence was seen as providing a balance between the needs of people and the needs of wildlife, something which several participants considered was currently lacking. This was exemplified by comments from Hector, another peat/compost producer, who said, 'the conservationists, they just take it too far, there's no balance left, I'm all for conservation and I love wildlife but also people need to make a living as well.' This balance is perhaps one of the most difficult areas for rewilding to negotiate when encountering, and countering, consumptive land uses, especially in negotiating attitudes to land and resources, management of land and resources, and the needs of people to make a living.

Rewilding's negotiation with farming in the Avalon Marshes

Grazing regimes, extensive farming, conventional agriculture and the negotiation between these and rewilding are significant aspects of conservation in the Avalon Marshes. There are two very important things to note here. First, some rewilding advocates take a purist view of rewilding, insisting that the involvement of any form or level of productive agriculture precludes something from being rewilding. Secondly, the grazing regimes at the Avalon Marshes are very much *conservation* grazing rather than *naturalistic* grazing.

As the name suggests, conservation grazing is predominantly a conservation technique wherein herbivores are used to maintain habitat in a particular condition (Hodder et al., 2005). This contrasts with naturalistic grazing, a rewilding strategy in which the process of grazing is important rather than the habitat it produces (Hodder et al., 2005). Naturalistic grazing occurs in Wild Ennerdale and is discussed in the next chapter, while conservation grazing and extensive farming are discussed here as part of the negotiations between rewilding and agriculture in the Avalon Marshes.

Several sections of the Avalon Marshes reserves are grazed by cattle, ponies and/or goats. Here, rewilding and extensive farming coexist, with Larry, a landowner/manager, explaining that livestock grazing performs a 'management' task which might otherwise be done with machinery. These words hint at how other-than-human animals occupy an interesting, and potentially problematic, role within rewilding. In one respect, by avoiding the use of machinery, other-than-human animals are ensuring that land is governed by 'natural processes' rather than 'human intervention', at least as far as possible – one has to overlook the fact that humans are responsible for the other-than-human animals being there. On the other hand, because these other-than-human animals are domestic and, ultimately, play a role in food production as well as performing their role in rewilding, they are not a truly 'natural' process but an artificial or cultural one. In this sense, the cattle, ponies and goats of the Avalon Marshes 'unsettle the modern division between the wild and the domestic' (Lorimer and Driessen, 2013). The way other-than-human animals involved in rewilding projects cross the boundary between 'wild' and 'not wild' requires humans to renegotiate their relationship with them and develop new modes of biopolitics, as will be discussed in greater detail in Part 2 (Cassidy, 2012; Lorimer and Driessen, 2013; Linnell et al., 2015; DeSilvey and Bartolini, 2018).

The multiple roles that other-than-human animals play in rewilding are inseparably imbricated and this was highlighted by participants. Maeve, a conservationist, talked about the way Exmoor ponies were deployed to 'tackle the scrubland' and 'keep that a suitable habitat for ground nesting birds'. Here, the ponies are embodying the role of human proxy (in tackling the scrubland), disturbance factor (in countering ecological succession and maintaining habitat for ground nesting birds), and analogues for extinct or extirpated species (having been chosen specifically for their morphological similarity, and possible genetic lineage to wild horses (Hovens and Rijkers, 2013; van Vuure, 2014)), all while performing the same activity. Comments from other participants hinted at how tensions might emerge in relation to the different biopolitical modes that apply to different roles. Niall, a conservationist, spoke about how cattle and goats were used for 'grazing' or 'scrub control' in the 'more difficult parts of the site' or 'the more scrubby units on the reserve' and highlighted that this was an onerous task because of the wet and difficult conditions which would be too harsh for less hardy animals. Octavia, another conservationist, expressed this even more clearly, saying that the cattle 'do a good job for us' (illustrating the cattle's role as proxies for humans) and going on to describe the risk of tick-borne disease that they are exposed to in the process. This illustrates how humans knowingly expose other-than-human animals to disease and harsh conditions when they enrol them in the rewilding of the Avalon Marshes in a way that they might not in other, more established modes of biopolitics such as those in conventional farming. This indicates that new modes of biopolitics are being created for the governance of the life, and possible death, of other-than-human animals involved in rewilding (this will be explored in more detail in Chapter 5).

With respect to how rewilding negotiates with farming more broadly, parts of the Avalon Marshes occupy land which was previously farmed and rewilding must therefore negotiate with farming's legacy. This is particularly evident where, as described by Alan, a conservationist, rewilding seeks to convert intensive farmland to wildflower-rich meadows. This requires the removal of nutrients from fertilizers added to the soil to suit the requirements of particular crops. Wildflowers are adapted to much lower nutrient levels and therefore cannot flourish in nutrient-enriched soil. The process of *un*improving soil involves a long and sometimes violent (*sensu* van Dooren, 2014) negotiation with the land. One of the best, or only, ways of reducing the nutrient content of the soil is to cut summer grass growth and remove the resultant hay from the land since leaving it lying on the ground allows nutrients to return to the soil. Similarly, livestock grazing allows for too much recycling of nutrients (in the animals' dung) to be effective at lowering nutrient levels. This was highlighted by Edward, a landowner/manager, who commented on the activities of the conservationists in managing the reserves: 'you don't have cattle on it [grassland] if you're going to take fertility out, they [the conservationists] should be mowing it and selling hay and selling hay and selling hay. But once you put animals in, it's going round in a circle'. Conservationists thus have to balance the, in this case undesirable, ability of grazing animals to recycle nutrients against the other roles they perform in the ecosystem.

The process of negotiating the nutrient content of the land left conservationists open to criticism from farmers, who could not understand why conservationists wanted to 'knock the goodness' out of the soil which farmers had spent time and money putting in. The language stakeholders used denoted the violence involved in the negotiation with soil (knock the goodness out of it) and also the view that conservationists 'don't know what they're doing', something which William, an archaeologist, felt was a commonly held belief by farmers:

> some farmers don't have a very good opinion of the farming ability of the nature conservation organisations. I think they still look at some of the land and think it's managed inefficiently, but then it's meeting nature conservation priorities isn't it? It's not out there to be maximum efficiency in terms of grazing.

This comment also alludes to a potentially irreconcilable difference between conservationists and farmers in that they have different priorities regarding how land should be managed for conservation versus agricultural production. Mutual compromise may, however, offer a way forward in the negotiation between farming and rewilding. For example, agroecological systems can offer mutual benefits for agriculture and conservation (as compared to more intensive agriculture systems of the present and recent past) and are gaining increased attention in illustrating that the goals of farming and conservation are not necessarily incompatible (Jackson, Maginnis and Sengupta, 2007; Neely and Hatfield, 2007; Thompson et al., 2007). This does, however, require compromise from rewilding in relation to one of its key tenets, that of reduction of human intervention. Such systems could, however, offer a means for rewilding and agriculture to negotiate from compatible, as opposed to conflicting, perspectives, offering a pragmatic solution to otherwise irreconcilable conflict between agriculture and rewilding and allow them to co-exist in a mutually beneficial way.

Rewilding in the Avalon Marshes also operates alongside ongoing as well as past agriculture and must, therefore, negotiate the physically constructed boundaries (e.g., ditches, fences, hedges), which demarcate the virtual boundaries of landownership. The other-than-human animals involved in rewilding will, however, often transgress these boundaries, going under, over or through them, or simply breaking them. For example, Edward, a landowner/manager, spoke about how the 'damn cattle got out' of one of the reserves of the Avalon Marshes and on to his land after breaking through fencing. Meanwhile, Larry, another landowner/manager, discussed the hypothetical reintroduction of large predators to England for whom the usual farm boundaries would be insignificant. Fencing sufficient to exclude such predators would, however, be prohibitively expensive for farmers, and Larry highlighted the consequences of not being able to exclude these predators from farmland:

> you've got someone living next door [to a rewilding project], farming that land, trying to make a living ... when he's lambing or calving in the spring, to that wild animal that's just like us going to a McDonald's, it's easy. ... they just go behind a sheep and pick a lamb up.

This highlights the need not only for rewilding to negotiate with farming but for humans to negotiate with other-than-human animals, including species for which we have, historically, had very little tolerance. For human tolerance of other species to improve, scholars suggest that we would need to hold a more enlightened position than we have in the past and change our approach to co-existing with other species (Wilson and Bruskotter, 2009; Arts, Fischer and van der Wal, 2012, 2016; Stohr, 2012). This is discussed extensively in Part 2 of this book which examines the other-than-human animals involved in rewilding in more detail, including species reintroductions and biopolitics in relation to these species.

Renegotiating rewilding

This chapter has explored the landscapes of the Avalon Marshes where soil and water are both extremely important. The peat soil is an integral part of the Avalon Marshes, forming the literal foundations of the landscape and also providing a more figurative foundation for the Avalon Marshes as they have evolved in the places and spaces where the peat industry used to be. Negotiating a position for itself is not easy for rewilding in the Avalon Marshes, it must compete in people's minds with all that the peat industry has been able to offer, and it must do so in a culture which has a history of staunch opposition to changes in landscape and the way it is managed. Agriculture is also an important land use and source of income in the Avalon Marshes, meaning that rewilding in the Avalon Marshes must negotiate with two competing forms of land use, both of which present unique challenges but also opportunities. Considerable resentment towards rewilding was expressed by stakeholders, with participants expressing frustration at what they saw as an imbalance between the needs of productive land use (in this case farming and peat production) and the needs of conservation. Nonetheless, rewilding in the Avalon Marshes continues, albeit under a different name and at a relatively small scale, biodiversity, or at least functional diversity, is increased and, as a result, ecological processes are taking place, often with human involvement. Thus, while human intervention is continuing at the Avalon Marshes, it is being reconfigured as interaction: human control is less dominant and there is more evidence of other-than-human agency exerting itself as humans and other-than-human animals renegotiate their coexistence.

Notes

1 Droves are old roads, originally used for driving livestock.
2 Active rewilding is an umbrella term used by scholars including Carver (2016) and Sandom et al. (2018) to describe a range of rewilding approaches all of which involve human intervention to a lesser or, more usually, greater extent.
3 This interpretation of creation of place is borrowed from Cloke and Jones (2001) who adopted Whatmore's (1999) interpretation of landscape as being the co-creation of place between multiple agencies to describe orchards.

References

Arts, K., Fischer, A. and van der Wal, R. (2012) 'Common stories of reintroduction: A discourse analysis of documents supporting animal reintroductions to Scotland', *Land Use Policy*, 29(4), pp. 911–920.
Arts, K., Fischer, A. and van der Wal, R. (2016) 'Boundaries of the wolf and the wild: A conceptual examination of the relationship between rewilding and animal reintroduction', *Restoration Ecology*, 24(1), pp. 27–34.
Carver, S. (2016) 'Rewilding... conservation and conflict', *Ecos*, 37(2), pp. 2–10.
Cassidy, A. (2019) *Vermin, Victims and Disease*. Cham: Palgrave Macmillan.
Cassidy, R. (2012) 'Lives with others: Climate change and human-animal relations', *Annual Review of Anthropology*, 41, pp. 21–36.

Clancy, C. and Ward, K. (2020) 'Auto-rewilding in post-industrial cities: The case of Inland Cormorants in Urban Britain', *Conservation and Society*, 18(2), pp. 126–136.
Cloke, P. and Jones, O. (2001) 'Dwelling, place, and landscape: An orchard in Somerset', *Environment and Planning*, 33, pp. 649–666.
DEFRA (2011) *The Natural Choice: Securing the Value of Nature*. London.
DeSilvey, C. and Bartolini, N. (2018) 'Where horses run free? Autonomy, temporality and rewilding in the Côa Valley, Portugal', *Transactions of the Institute of British Geographers*, 44, pp. 94–109.
Gammon, A. R. (2018) 'The many meanings of rewilding: An introduction and the case for a broad conceptualisation', *Environmental Values*, 27(4), pp. 331–350.
Gillson, L., Ladle, R. J. and Araújo, M. B. (2011) 'Baselines, patterns and process' in R. J. Ladle and R. J. Whittake (Eds) *Conservation Biogeography*. Chichester: Blackwell.
Höchtl, F., Lehringer, S. and Konold, W. (2005) '"Wilderness": What it means when it becomes a reality – A case study from the southwestern Alps', *Landscape and Urban Planning*, 70(1–2), pp. 85–95.
Hodder, K. H. et al. (2005) *Large Herbivores in the Wildwood and in Modern Naturalistic Grazing Systems*. Peterborough. Available at: publications.naturalengland.org.uk/file/93025
Hovens, H. and Rijkers, T. (2013) 'On the origins of the Exmoor pony: Did the wild horse survive in Britain?', *Lutura*, 56(2), pp. 129–136. Available at: http://www.zoogdiervereniging.nl/sites/default/files/imce/nieuwesite/0verigen/downloads/Lutra_Text_56_2_lowresv2.pdf#page=57
IUCN CEM (2021) *Rewilding Principles*. Gland: IUCN. Available at: https://iucn.org/sites/default/files/2022-10/principles_of_rewilding_cem_rtg.pdf
Jackson, W., Maginnis, S. and Sengupta, S. (2007) 'Planning at landscape scale' in S. Scherr and J. McNeely (Eds) *Farming with Nature: The Science and Practice of Ecoagriculture*. Washington: Island Press, pp. 308–321.
Jørgensen, D. (2015) 'Rethinking rewilding', *Geoforum*, 65, pp. 482–488.
Katz, C. (1998) 'Whose nature, whose culture? Private Productions of space and the "preservation" of nature' in B. Braun and N. Castree (Eds) *Remaking Reality: Nature at the Millennium*. London: Routledge, pp. 46–63.
Lasanta, T., Nadal-Romero, E. and Arnáez, J. (2015) 'Managing abandoned farmland to control the impact of re-vegetation on the environment. The state of the art in Europe', *Environmental Science and Policy*, 52, pp. 99–109.
Lennox, G. D., Dallimer, M. and Armsworth, P. R. (2012) 'Landowners' ability to leverage in negotiations over habitat conservation', *Theoretical Ecology*, 5, pp. 115–128.
Linnell, J. D. C. et al. (2015) 'Framing the relationship between people and nature in the context of European conservation', *Conservation Biology*, 29(4), pp. 978–985.
Lorimer, J. and Driessen, C. (2013) 'Bovine biopolitics and the promise of monsters in the rewilding of Heck cattle', *Geoforum*, 48, pp. 249–259.
Lorimer, J. et al. (2015) 'Rewilding: Science, practice, and politics', *Annual Review of Environment and Resources*, 40, pp. 39–62.
Macdonald, B. (2019) *Rebirding: Rewilding Britain and Its Birds*. Exeter: Pelagic Publishing.
Matless, D. (2014) *In the Nature of Landscape*. Chichester: Wiley Blackwell.
McEachern, C. (1992) 'Farmers and conservation: Conflict and accommodation in farming politics', *Journal of Rural Studies*, 8(2), pp. 159–171.
MHCLG (2012) *National Planning Policy Framework*. London.
MHCLG (2019) *National Planning Policy Framework*. London.
Moss, S. (2016) *Wild Kingdom: Bringing Back Britain's Wildlife*. London: Square Peg.
Murray, M. (2017) 'Wild pathways of inclusive conservation', *Biological Conservation*, 214, pp. 206–212.

Navarro, L. M. and Pereira, H. M. (2012) 'Rewilding abandoned landscapes in Europe', *Ecosystems*, 15(6), pp. 900–912.
Neely, C. and Hatfield, R. (2007) 'Livestock Systems' in S. Scherr and J. McNeely (Eds) *Farming with Nature: The Science and Practice of Ecoagriculture*. Washington: Island Press, pp. 121–142.
Nogués-Bravo, D. et al. (2016) 'Rewilding is the new pandora's box in conservation', *Current Biology*, 26(3), pp. 87–91.
Pettorelli, N. et al. (2018) 'Making rewilding fit for policy', *Journal of Applied Ecology*, 55 pp. 1114–1125.
Sandom, C. J. et al. (2018) 'Rewilding in the English uplands: Policy and practice', *Journal of Applied Ecology*, 56 pp. 266–273.
SCC (2013) *Minerals Topic Paper 3: Peat Reserves and Supply*. Taunton.
SCC (2015) *Somerset Minerals Plan*. Taunton.
Stohr, W. G. (2012) 'Trophic cascades and private property: The challenges of a regulatory balancing act and lessons the UK can learn from the reintroduction of the American Gray Wolf', *University of Baltimore Journal of Land and Development*, 2(1), pp. 15–51.
Taylor, P. (2017) 'Hard Brexit – Soft rewilding?', *Ecos*, 38(1).
Thompson, J. et al. (2007) 'Biodiversity in agroecosystems' in S. Scherr and J. McNeely (Eds) *Farming with Nature: The Science and Practice of Ecoagriculture*. Washington: Island Press, pp. 46–60.
van Dooren, T. (2014) *Flight Ways – Life and Loss at the Edge of Extinction, Nucleic Acids Research*. New York: Columbia University Press.
van Vuure, C. (2014) 'On the origin of the Polish Konik and its relation to Dutch nature management', *Lutra*, 57(2), pp. 111–130.
Whatmore, S. (1999) 'Hybrid geographies: Rethinking the "Human" in human geography' in D. Massey, J. Allen and P. Sarre (Eds) *Human Geography Today*. Cambridge: Polity Press, pp. 22–40.
Wilson, R. S. and Bruskotter, J. T. (2009) 'Assessing the impact of decision frame and existing attitudes on support for wolf restoration in the United States', *Human Dimensions of Wildlife*, 14(5), pp. 353–365.

3 Wilding the Ennerdale Valley

The Wild Ennerdale landscape

Wild Ennerdale reveals itself only slowly, glimpses of its fells are caught and lost with the rises and falls of the road as the driver approaches. Ennerdale Water, the valley's eponymous lake, remains hidden until the last moment when it and the visitor emerge from the trees simultaneously, and its most famous peaks, Pillar Rock, Haystacks and Great Gable, reveal themselves only to the walker who ventures into the upper valley. On encountering Wild Ennerdale from the Bleach Green entrance and leaving the metalled car park behind, the walker soon reaches Ennerdale Water, lapping against its man-made weir and revetments. There is a choice of paths around the lake – the way to the left is a wide, well-formed track which follows the lake shore to the Bowness Knott entrance where the visitor can join the forest road all the way to the head of the lake. The way to the right offers the walker a very different journey: starting gently enough through a grassy meadow, the trail quickly deteriorates to a single file track before petering out altogether as it, and the walker, scramble up and over Angler's Crag. The route becomes increasingly entangled with nature as it continues towards the head of the lake, with the path often indistinguishable from the streams that run beside, over and sometimes along it. Whichever way is chosen, the routes converge on the banks of the River Liza, the river which feeds Ennerdale Water. The Liza meanders and braids its way through the valley, untrammelled by any human hand. Following the Liza leads the visitor further from civilization and into the upper valley, where the fells climb higher and the plantation conifers start to give way, either to moorland or to deciduous saplings, native to the valley but introduced by human agency. This entanglement of the 'natural' and the 'artificial' is characteristic of Wild Ennerdale, its rewilding seeking to reduce human intervention yet using humans to catalyse the process of natural regeneration.

This chapter discusses the rewilding, or 'wilding', of the Ennerdale Valley – the catchment of the River Liza and Ennerdale Water in England's Lake District in the northwestern county of Cumbria. It introduces the landscapes of Wild Ennerdale and discusses themes which emerged during interviews with rewilding's stakeholders and practitioners. Key among those stakeholders are

DOI: 10.4324/9781003456490-4

the Wild Ennerdale partners: Forestry England (a government agency which manages public woodland), Natural England (a public body to advise the government and protect the environment), the National Trust (a heritage conservation charity) and United Utilities (a water company). Together, these partners have developed a stewardship plan for the land and water they collectively own. This landownership corresponds almost, but not entirely, with the catchment of the Ennerdale Valley (three comparatively small parcels of land within the Ennerdale Valley catchment are privately owned and are therefore managed independently from the partnership). It is the land under the management of the Wild Ennerdale Partnership (some 4300 hectares) which forms the focus of this chapter.

The naming of (re)wilding in the Ennerdale Valley

The practitioners involved in the stewardship, as they term it, of Wild Ennerdale are scarcely more comfortable with the term rewilding than those undertaking similar work in the Avalon Marshes. Wild Ennerdale's first stewardship plan took effect in early 2006 but, according to that same plan, was the culmination of five years of discussions, with a memorandum of understanding using the appellation 'Wild Ennerdale' having been signed in 2002 (Wild Ennerdale, 2006). Given that Wild Ennerdale adopted its title at least as early as 2002, the employment of it predates rewilding as a concept gaining significant currency within media discourse in the UK, something which happened circa 2009 (Jepson, 2016). From this perspective then, the adoption of the name Wild Ennerdale preceded the term rewilding becoming fashionable and in turn controversial, something which did not happen until after 2013 when Monbiot's (2013) *Feral* further popularised and subsequently problematised the term (Jørgensen, 2015; Jepson, 2016; Prior and Brady, 2017; Tanasescu, 2017; Deary and Warren, 2018; Gammon, 2018; Pettorelli et al., 2018; Sandom et al., 2018).

The controversy has resulted in the Wild Ennerdale partners being at pains to distance themselves from the term rewilding, insisting that the naming of Wild Ennerdale predated rewilding's popularity (or *un*popularity) as a term, and emphasising their reluctance to rename the partnership now either to associate themselves with or dissociate themselves from the rewilding movement. Brian, a forester, made this explicit saying 'we've steered clear of using it [rewilding] partly because we'd developed our own language beforehand, before rewilding became a term you heard in the UK, we were using terms like 'sense of wildness' to describe what you felt like to be in a wild place'. Callum, a ranger, made a similar point, emphasising the fact that Wild Ennerdale was named independently of the discourse surrounding rewilding, instead arising 'organically' from the sense that people get from being in Ennerdale Valley: 'it wasn't necessarily aimed at being a rewilding place at the time, that more came organically, the label of "wild", and anyone who's spent some time in the valley can probably totally understand that'. That the concept of 'wild' is closely associated with the Ennerdale Valley is something Ross, a conservationist, drew attention to: 'the premise for calling it 'Wild Ennerdale' was that "wild" was just the collective feeling we had, that the valley gave us all when we talked about it'. Ross went on to highlight that the lack of suitable alternative words is a recurring issue when talking about wildness or rewilding: 'natural processes led management decision making Ennerdale' is not particularly easy to trip off the tongue is it?' While the example they use is something of a *reductio ad absurdum*, the comment certainly echoes scholars who point to the difficulty of finding a word other than rewilding which conveys the same meaning while also evoking similar feelings of inspiration and optimism (Carver, 2016). This presents a difficulty for those who might prefer not to use the term rewilding, but do not feel that there is an appropriate synonym.

Practitioners were highly aware that the term rewilding has generated significant interest in the years since Wild Ennerdale was named, with Ross saying that rewilding has entered 'far more into the public conscience'. Max, a conservation advisor, described how, while 'maybe ten years ago, they would have described themselves as rewilding' that they would not do so now and said that the current reluctance stemmed from '*Feral*, from George Monbiot, and some of that reluctance has just come more recently with the media getting interested and there being maybe again more polarised debate about what the uplands are for or what land's for'. Indeed, Max was extremely reluctant to use the term rewilding, saying that the word had 'lost its definition', become 'toxic' (echoing the words of Sandom et al. (2018)), become inextricably associated with 'debate about predators', is perceived to 'exclude people rather than include people', and 'doesn't help in any way explain what we're trying to do'. For this reason they said that there was 'no value in pursuing what it does mean' and that rewilding is 'just not worth using' 'because we've got other ways of describing what we're trying to do that maybe are more easy for people to come on board with, and don't have the toxicity associated with them'. Max went so far as to claim that 'none of us would describe ourselves as rewilding' because it 'just gets people off the wrong way' and stated that they instead describe themselves as 'a natural processes project, where we're letting natural processes take more of a hand in determining the outcomes'. Brendan, a conservationist, made a similar point, although not so forcefully, saying that 'we tend not to use the term "rewilding" I guess because we never have done, it's not something that we feel we've needed to change our name to adapt to'. This point is similar to that made by practitioners in the Avalon Marshes who were reluctant to change the name of their project to involve themselves in a movement which they preceded.

Like the practitioners quoted above, stakeholders also had concerns over the meaning of the term rewilding, mirroring the debate among scholars. Tristan, a policy officer, described rewilding as 'not meaning anything', 'in danger of being hijacked' and 'meaning too many different things to too many different people'; this resonates with the work of scholars who note rewilding's lack of an agreed definition (something which has changed only recently) (Prior and Brady, 2017; Hayward et al., 2019) and describe the term as plastic (Jørgensen, 2015). For this reason, Tristan saw it as more useful to 'tell people what you're actually doing, as opposed to using this concept [of rewilding]'. Meanwhile, other stakeholders discussed whether Wild Ennerdale *was* wild, and therefore whether the term rewilding was applicable. Naomi, a parish council representative, argued that Ennerdale is already wild, so 'what do you mean by rewilding?'. Conversely, Peggy, a landowner/manager, argued not only that Wild Ennerdale is *not* currently wild but also the improbability that 'anything will end up as wild, so rewilding isn't a true definition'. Peggy noted, however, the positive connotations that rewilding can have (echoing scholars who describes rewilding as 'inspiring', e.g., Sandom et al. (2013), Arts, Fischer and van der Wal (2016) and Sandom and Wynne-Jones (2019) and that it is therefore 'a word that's used for effect, rather than for a true description'. Similarly, Hannah, a business owner/manager, talked about the term's value for 'marketing' purposes, especially in terms of getting people 'interested' in what's happening. From the opposite perspective, other stakeholders saw the implications of the term rewilding as wholly negative, saying that 'using words like rewilding doesn't help' and that the word rewilding 'gets people wound up'. These stakeholders related the negative connotations of rewilding to its association with the reintroduction of large mammals, particularly large predators (especially lynx and wolves), and the consternation related to this.

Stakeholders also questioned whether the activity in Ennerdale could really be considered rewilding asking 'isn't it just land management? Isn't it just land management to improve what it is?'. In particular, stakeholders highlighted the contradictions inherent in the definitions and praxis of rewilding, questioning the extent to which rewilding can or cannot involve management. Stephen, a landowner/manager, was particularly vociferous on this point, first highlighting the 'contradiction' inherent in the management of roe deer in the valley, with 'stalkers in there blowing their heads off all over the place', while the Wild Ennerdale partnership is purporting to be reducing human intervention. Stephen saw this approach to management as inconsistent, particularly in relation to whether things are 'wild' or 'natural' (reminiscent of debates over the distinction between nature and culture and whether humans are part of or apart from nature (e.g., McKibben, 1990; White, 1995; Bowker, 2000; Nye, 2000; Helmreich, 2005; Carver, 2007; Brown, McMorran and Price, 2011; Cassidy, 2012; DeMello, 2012; Seddon et al., 2014; Head, 2015; Saunders, 2016)):

> the whole principle of rewilding it would appear requires the use of lots and lots of volunteer hours to make it properly wild so where do we draw the line between 'wild' and 'management'? If we've got loads of people in there removing undesirable species and replacing them with desirable species, that's not rewilding. That's not natural processes. That's management. That's farming or forestry. That is doing something to deliver an outcome. That is positive commercial management.

Adam, another landowner/manager, was equally convinced that, strictly speaking, rewilding should not involve management saying, 'to me rewilding means just let it grow on its own, don't manage it in any way whatsoever, no management at all'. This is not to suggest, however, that Adam thought this was a good approach – he went on to say that 'eventually what you'd end up with is really a lot of rubbish if you didn't manage it, there's got to be some kind of management, there's got to be human intervention'. This illuminates a difficult balance for rewilding to negotiate – criticised on one hand for betraying its principles if human intervention is involved and, on the other, for resulting in landscape deterioration *without* human involvement.

The practice of rewilding in the Ennerdale Valley

Despite the concerns of Wild Ennerdale's practitioners regarding the term rewilding, Wild Ennerdale is, nonetheless, part of the European Rewilding Network, thereby aligning and allying itself with rewilding. This is announced on Wild Ennerdale's website with the statement: 'Wild Ennerdale is excited to be a member of the European Rewilding Network' (Wild Ennerdale, 2019). The decision to join the network was justified by Vanessa, a landowner/manager, who explained that it enhanced their ability to visit other rewilding sites, thereby growing their network and facilitating knowledge exchange. Furthermore, contrary to the assertion by Max, quoted earlier, that 'none of us would describe ourselves as rewilding' other practitioners *did* refer to Wild Ennerdale as rewilding, albeit sometimes to admit the ambiguity of the term with Callum, a ranger, saying 'I think it's quite a broad stroke approach that Wild Ennerdale takes to rewilding and probably a lot of people would look at it and dispute whether it is truly rewilding or not, you could definitely argue as to whether this is rewilding or if it isn't rewilding'. Meanwhile, Vanessa, said that she saw rewilding as 'allowing natural processes to lead for the benefit of nature and people' and emphasised that this was the interpretation which Wild Ennerdale used in its stewardship plan. Indeed, Vanessa was deliberately invoking the Wild Ennerdale vision, stated in its stewardship plan and on its website, 'to allow the evolution of Ennerdale as a wild valley for the benefit of people, relying more on natural processes to shape its landscape and ecology' (Wild Ennerdale, 2006, 2019). This comment illustrates how, together with associating itself with the term rewilding (notably by joining the European Rewilding Network), Wild Ennerdale also has other aspirations which demonstrate family resemblance to rewilding (i.e., it seeks to restore ecological functioning) and this is demonstrated in its current and proposed land management.

Significant changes to land management in Wild Ennerdale have already been instigated and further changes are proposed. Perhaps the three most significant changes which are underway are in woodland management, grazing regimes and boundary maintenance (specifically of drystone walls). With regard to woodland management, Forestry England is moving away from commercial forestry in the Ennerdale Valley. In place of Sitka spruce monocultures, woodland creation with native trees is underway both by tree planting and by providing a seed source to enable natural regeneration. The rationale behind providing a seed source was explained by Max, a conservation advisor:

> we've done a whole load of planting in that upper end [of the valley], hundreds of thousands of trees, not to create a wood, but to create a seed source, because actually we're interested in the process, not the outcome, so the process is to say 'right here's a load of seeds we're throwing into the system, away you go woodland, regenerate yourself'

This focus on reinstating natural processes, and thereby restoring ecosystem function, supports the case for Wild Ennerdale being an example of rewilding.

The emphasis on restoring ecosystem function is also evident in the changes in grazing regimes. Sheep farming is being displaced (though not entirely replaced) by naturalistic grazing with native cattle in which the emphasis is on grazing as a natural process rather than the product of that grazing (e.g., creating or maintaining a certain habitat or ecosystem, or indeed in meat production as is the case with sheep farming) (Hodder et al., 2005). The other focus in naturalistic grazing is on the species involved, with it being usual for there to be an emphasis on them being native and on their function as analogues for absent species. This is discussed in more detail later in this chapter, it is, however, useful to highlight here that the emphasis from Wild Ennerdale practitioners regarding the cattle was that they 'create regeneration niches, because they're a bit heavier [than sheep] so their footprints become regeneration niches, they don't compact the soil'; this poaching of the ground created by the cattle's hooves is 'the disturbance the [Wild Ennerdale] partnership look for, pushing seeds into the ground, helping them germinate', in other words restoring a more naturalistic grazing process also restores other natural processes.

Another change underway regarding the maintenance of boundaries, including the repair of drystone walls, relates to sheep and cattle management in the valley and also to farming more broadly. As part of the wilding of the Ennerdale Valley, the Wild Ennerdale partnership is interested in 'blurring' or softening hard boundaries within the valley. This is being done by removing fences and by allowing drystone walls to 'gently fall down', as Vanessa, a landowner/manager put it. Stephen, another landowner/manager, described this as allowing 'the woodland and the fell ground to run together' (i.e., the effect the Wild Ennerdale Partnership is hoping to achieve), which is appropriate for naturalistic grazing with cattle but presents problems for traditional sheep farming. Again, this is discussed in more detail later in this chapter.

The ongoing changes in management of the Ennerdale Valley, with its focus on increasing biodiversity, ecological functioning and natural agency, together with a reduction in human intervention and the fact that it operates at large scale and associates itself with the term rewilding, positions Wild Ennerdale within the 'nature led, active culture' segment of the rewilding typology explained in Chapter 1; Wild Ennerdale encourages natural, ecological processes to resume and/or continue, with human intervention to facilitate, accelerate or moderate this process. Given that Wild Ennerdale is, in every practical sense, a case of rewilding, it is useful to examine the way it is perceived as such.

The perception of rewilding in the Ennerdale Valley

A dominant theme in the way stakeholders perceived the rewilding of the Ennerdale Valley was that of abandonment. This is perhaps unsurprising given that land abandonment in and of itself is one interpretation of rewilding and can also be a tenet of other interpretations (Höchtl, Lehringer and Konold, 2005; Navarro and Pereira, 2012; Jørgensen, 2015). What was of note in stakeholder responses, however, was the negative attitude to abandonment, associating it with connotations of deserting or forsaking land rather than the sense of relinquishing control to another (in this case the other being nature) or even the sense of to release, set free or liberate, the prefix 'a' making it the negation of 'bandon' being 'dominion' or 'control'. Wesley, an environmental farming advisor, explained that this was true with regard to perceptions of rewilding generally ('when you think of rewilding, you basically think of just land abandonment') and the rewilding of Ennerdale specifically, giving the example of a group of farming students who were taken to visit Wild Ennerdale and their perception being that 'what they saw was just tantamount to land abandonment'.

Stephen, a landowner/manager, was even more specific, drawing attention to Low Moor End Farm, a farm at the entrance to the Ennerdale Valley. Low Moor End Farm was purchased by United Utilities as part of compensatory measures they were required to put in place as a result of the adverse effects their activities had on the River Ehen Special Area of Conservation (United Utilities, 2020). Stephen was clear both that he considered the farm abandoned and that he considered abandonment to be rewilding, saying that United Utilities 'bought a little farm in the bottom of the valley, just outside the village, and they have abandoned that, that is what I would consider to be rewilding'. He noted, however, that 'we haven't to use the word 'abandon', because the Wild Ennerdale people get really, really angry if we describe it as abandonment, even though they've left the house empty and it's falling down'. In his view, this was because the practitioners at Wild Ennerdale made a distinction between 'abandonment' and 'rewilding', perhaps to avoid the negative connotations of abandonment, although by replacing abandonment with rewilding they are perhaps stepping out of the linguistic frying pan and into the linguistic fire.

Clare, a conservationist, recognised that stakeholders saw Low Moor End Farm as abandoned but attributed this to stakeholders' lack of a 'long-term view' of the process of rewilding and also to the way the appearance of the farm offended their aesthetic sensibilities:

> this land is a bit of a bone of contention with some of the locals because they look at this and think it's abandoned and it's awful. I can understand what people say, it does look abandoned and all the rest of it, but it's that long-term view that you have to take with these things.

Stakeholders, however, gave different, less superficial reasons than aesthetic qualities for their concern over abandonment. First was the perception that rewilding resulted in the cultural element of Wild Ennerdale's landscape being disregarded and devalued, potentially resulting in the loss of that landscape. Second was the concern that the rewilding of one area simply transposed environmental problems to another area (this echoes criticism of rewilding by scholars that rewilding outsources food production, and the associated environmental degradation, to other countries so that rewilding can occur in its host country, e.g., Navarro and Pereira, 2012; Fairlie, 2013). Duncan, a ranger, articulated the perceived lack of appreciation for the value of the Lake District's heritage landscape when they said 'people look at this landscape and think generations of farmers have farmed it, they've removed all the rocks, there's a lot of blood and sweat gone into it to make the land productive and what they see UU [United Utilities] doing is letting it go to waste'. This comment echoes scholars who discuss the way cultivation represents progress, mastery over nature, and the result of many generations of labour (Ingold, 2000; Carver, 2007; Tsing, 2012). Indeed, Karl, a business owner/manager, highlighted the risk that the legacy of those generations of labour could be lost and that the opportunity to continue the farming tradition is being taken away by rewilding, saying of Low Moor End Farm that it had 'been a working farm for hundreds of years but because they're rewilding that, they're taking the opportunity away'.

The transposition of environmental problems from one place to another was also expressed by Karl, saying that the abandonment of the farmhouse at Low Moor End meant that:

> now they have to build another house because that house doesn't exist anymore, so, it's a mad, mad way of rewilding because all you're doing is the neighbouring town now has to build more houses and expand into their wilder areas because you're trying to take the human factor out of a wild place.

A similar point was made by Stephen, a landowner/manager, although he focused more on the human rather than the environmental impact of abandoning land saying that 'there's kids in the village [Ennerdale Bridge] that have to move to Cleator Moor[1] because they can't get a house'. This comment hints at the rural depopulation which rewilding has been suspected of causing (Jones and Comfort, 2020), a subject which is entangled with the loss of farming and the damage to rural communities and which was expanded on by Wesley, an environmental farming advisor. His comments related not to Low Moor End Farm specifically but to the rewilding of farms generally, which they saw as 'massively controversial' because of the 'issue of farms for the next generation' given that, if the tenancy on a National Trust farm becomes available the 'National Trust say no, our policy is we're going to rewild that and rent out that building as a holiday cottage or rent it out on a commercial basis, that doesn't speak well to farmers because it's just another blocker in that succession'. The succession which this stakeholder is concerned about is the succession of older farmers by the younger generation and the difficulty for young farmers entering the business, removing farms from productive agriculture for rewilding purposes is seen by stakeholders as exacerbating the difficultly of farm succession.

Rural depopulation is another significant concern in relation to rewilding. While scholars note that concern exists in relation to rewilding *causing* rural depopulation (e.g., Sandom and Wynne-Jones, 2019; Jones and Comfort, 2020), a much more common theme is of rewilding occurring where rural depopulation and land abandonment have already taken place (e.g., Navarro and Pereira, 2012; Ceausu et al., 2015; Lorimer et al., 2015; Navarro and Pereira, 2015; Jepson, 2016; van der Zanden et al., 2017; DeSilvey and Bartolini, 2018; Jones and Comfort, 2020). It is important to note, however, that these scholars focus heavily on rural depopulation and subsequent land abandonment in continental Europe as opposed to Britain. In Britain, rural land values are higher than in many other European countries, certainly as compared to Bulgaria, Croatia and Romania, for example, which have some of the lowest land prices and are home to several of Rewilding Europe's most ambitious projects (Eurostat, 2018; Rewilding Europe, 2020). High land prices in Britain reduce the likelihood of land abandonment, thus limiting the opportunity for rewilding to *follow* rural depopulation but raising the possibility of *inducing* it. This is a cause for considerable concern in Britain, where such abandonment has the potential to cause significant changes to valued heritage landscapes.

The negotiation of rewilding in the Ennerdale Valley

Rewilding's negotiations with culture

The Lake District is one such heritage landscape, and being part of the Lake District National Park means that Ennerdale shares the associations which the Lake District more broadly brings to mind. The Lake District conjures up associations with the Romantic poets, especially Wordsworth, calling to mind images of shepherding, hill sheep farming and Herdwick sheep, recently captured by James Rebanks in his book *A Shepherd's Life* (2016), and evokes memories of, or longings for, fell walking. Indeed, since the Lake District exists so powerfully in the collective imagination, holding 'all the loved places that the exile longs for' (Thompson, 1946, p. 3), Wild Ennerdale is, to some extent, present in our minds without our being there, perhaps without our ever going there, as a place we plan to visit, or just imagine we might visit. Vanessa, a landowner/manager, attributed this sense to England's cultural heritage, specifically the Lake Poets:

> the Romantic poets probably are at the heart of it because everyone knows Wordsworth's 'Daffodils' and everyone sees pictures of Ullswater or wherever it was supposed to be where that poem was written, I think anyone who's ever come here or has read about it or seen it just seems to feel a kind of connection, people who don't come to special places still feel quite a connection with them because it's that feeling of well if they did go, this is probably how I would feel about it.

This might be considered a deep sense of familiarity, in that Wild Ennerdale feels familiar at some subconscious level, even if we have never been there, in a similar way to which it has been suggested that an often-visited landscape can feel familiar even if its appearance is always changing (Cloke and Jones, 2001). To some extent, those acquainted with the Lake District, either through direct or indirect experience, carry in their minds, and more or less consciously engage with, imaginative constructions of sheep, fells, shepherding, the Lakes, Cumbria, England and the English countryside (Cloke and Jones, 2001). If they subsequently visit the Lake District, these imaginative constructions combine with the receipts of their senses to create 'complex sensory and imaginative, dynamic, collages of being' in the Lake District (Cloke and Jones, 2001, p. 663).

Because of this very special place which the Lake District occupies in the English public imagination, it presents an extremely complex landscape with which rewilding interfaces. Not only is the landscape one which is collectively recognised and appreciated, but it is one which has been created by a long interaction between humans and their environment (Linnell et al., 2015). Indeed it has been suggested that 'what we really treasure [about the Lake District] is the combination of the wild and the cultivated which seems to have reached perfection' (Thompson, 1946, p. 6), that it is the kind of 'perfect symbiosis between nature and human management' (Lasanta, Nadal-Romero and Arnáez, 2015).

Changes to this landscape are, therefore, highly likely to cause consternation and even controversy among those who value the landscape in its current form. This is particularly true with respect to rewilding which disrupts the existing interaction between people and nature and which has the potential to change the landscape in ways that are radical and unpredictable (Carver, 2007; Arts, Fischer and van der Wal, 2012; Lorimer and Driessen, 2013; Seddon et al., 2014; Sandom and Macdonald, 2015; Jepson, 2016; Svenning et al., 2016; Drenthen, 2018; Gammon, 2018; Pettorelli et al., 2018; Root-Bernstein, Gooden and Boyes, 2018; Wynne-Jones, Strouts and Holmes, 2018; Jepson and Blythe, 2020), and will therefore almost certainly result in a landscape which is different from the 'traditional' Lake District landscape. This, unsurprisingly, was a dominant theme which emerged from interviews, with participants stating how much people value the Lake District landscape in its current form, how people are resistant to change generally and how people are particularly resistant to the changes associated with rewilding because it results in a landscape which is 'not traditional' and is 'messy'[2]. The discussion of mess echoes scholars who acknowledge that rewilding uses 'practices that encourage less conventionally "pretty" or "beautiful" landscapes' and that because there is less management the preservation of 'aesthetically-valued qualities' or the avoidance of 'difficult aesthetic experiences' will not be possible (Prior and Brady, 2017; also Tree, 2018). Scholars suggest that management, at least in the short term, to avoid these difficult aesthetic experiences and create swift and positive landscape change can help to negotiate these difficult 'teething periods' of rewilding (Deary and Warren, 2018).

The first of these concerns was evident in a comment from Callum, a ranger, who talked about how much people value what might be described as the classic Lake District view; 'you see across the tops of the fells and it's a beautiful thing, everyone who comes to the Lake District loves that, standing on top of a fell and looking out into the distance and being able to see all the landscape around you'. This statement is supported by the fact that the view of Wastwater (one of the Lakes District's most iconic lakes) was voted 'Britain's favourite view' in 2007 (Visit Cumbria, 2020). Callum went on to acknowledge that this 'classic' view, which is valued so highly by the British public, is 'a manufactured landscape, it's not entirely natural' and that 'the Lake District we all know and love is a product of the way that humans have used the land'. This reflects the discussion in Chapter 1 of 'natural' versus 'cultural' landscapes and the role of humans in nature. Callum's comment clearly places humans as *apart from* rather than *a part of* nature and re-emphasises the needs for rewilding to negotiate this relationship.

Regardless of whether the landscape is 'natural' or 'cultural', however, it is, undeniably, highly valued and people are, therefore, very reluctant to see it change, as evinced by quotes from other practitioners. Vanessa, a landowner/manager, recognised that 'people feel very, very passionately about it [the landscape] and some people don't want to see it change at all', and Brendan, a conservationist, said that:

> people are super cautious about change, and people generally, when you're talking about a landscape, get very nervous when you start talking about changing things. Thinking back to when we started with Wild Ennerdale, I can remember comments like, 'why are you messing with my back garden, it's beautiful as it is', there's this fear that it's going to change for the worse.

This attitude was countered by comments such as those from Ross, another conservationist, who contended that changes to the Lake District landscape as a result of rewilding would alter but not diminish it, saying 'the views just change, you get views that are framed between the trees and they can be just as impressive'. Ross went on to say that 'to me it doesn't look messy' but acknowledged that 'to an awful lot of people it does'. The attitude which Ross described was indeed expressed in comments from stakeholders like Duncan, a ranger, who called for management of the Lake District generally and the Ennerdale Valley specifically to avoid just this sort of mess but also, perhaps more significantly, because they felt the need for human management of 'wild' places saying 'it's never going to be self-regulating, it's always going to need management, if it was left to self-regulate it would all in my view just turn to scrub, for me you have to manage wild places in this country'. Duncan went on to explain that this was because 'Wild Ennerdale is very much what people perceive as wild, which isn't necessarily an actual wild landscape, it's an attractive landscape, so I think there'll always be a role for organisations to manage the land'. This comment echoes statements made 80 years ago and yet which seem as relevant as ever: 'if we were to leave the Lake District entirely to nature it would soon become disorderly – unpleasantly disorderly by civilized standards' (Thompson, 1946, p. 15), highlighting our highly cultural perception of nature and our anthropocentric view that nature 'needs' human management.

An interview with Naomi, a parish council representative, provided a particularly clear illustration of the negotiation between rewilding and an aesthetically pleasing landscape. The interview was conducted while walking along the western edge of Ennerdale Water. *En route*, Naomi outlined the possibility of changes to the lake shore as a result of the lake's decommissioning as a reservoir. In the future this may result in the removal of the revetments and weir which currently contain the lake and regulate its level. If they are removed, Naomi suggested that 'the water level will drop quite a lot in the lake, so it'd totally change this end of the lake just because it would become a wetland habitat' but 'in the first years it would probably just be some mud flats'. Mudflats would present a very different and potentially *difficult* aesthetic experience to the one currently enjoyed (Prior and Brady, 2017).

Changes will also arise from a reduction in sheep farming, but such changes have the potential for deeper cultural resonances than simply changing a view or lake shore. Sheep farming has been part of the Lake District landscape for centuries, and the way that sheep are 'hefted' onto the fells is a remarkable achievement combining aspects of landscape, shepherding and ovine instinct. Hefting uses the sheep's tendency to form strong territorial attachments to a place to bond flocks to a particular area, allowing them to be farmed without fences and has been called a human–animal–place relationship (Gray, 2014). Rewilding has the potential to destabilise this relationship because hefting is not inviolable, it relies on a long-standing relationship with the land, and generations of sheep must be kept in the same area so that place attachment can be passed down from one generation to the next in order for the hefting to be maintained (Gray, 2014; Armstrong, 2016). Since rewilding is reducing sheep farming in the Ennerdale Valley, it is disrupting hefting in the area and could potentially result in permanent changes to the landscape as, not only is the heritage landscape lost, but the 'management techniques required for its conservation' are also lost, something which has been identified in other rewilding areas (Lasanta, Nadal-Romero and Arnáez, 2015, p. 101). Adam, a landowner/manager, described the cultural threat to the Ennerdale Valley (and by extension the wider Lakes District) posed by rewilding as 'criminal', explaining that the sheep are 'not just bought and fired out there [on to the fells], they've been with that area of land for potentially hundreds if not thousands of years, they've been there for a long, long time'. This (real or perceived) threat to farming is discussed in more detail in the next section.

Rewilding's negotiation with farming

When, in 2013, George Monbiot, now (in)famously, described the British uplands as 'sheepwrecked' by farming he created, according to Ross, a conservationist, 'a bit of a ding dong' in the Lake District, traces of which were still evident in interviews with stakeholders and practitioners. On one hand, stakeholders like Stephen, a landowner/manager, argued not only that sheep farming is 'an integral part of the whole of lake District management system' but also that it can be done in a way that has a positive impact on biodiversity: 'put a few more sheep in there, recreate the balance and everybody will be happy and your grass will be better, you'll have more biodiversity'. Meanwhile, practitioners like Max, a conservationist, argued that, as a result of sheep farming, the Lake District fells were 'very degraded, very impoverished, very uniform, poor structure, low biodiversity, low carbon, ecologically it's falling apart, it's eroding, we're losing soil, we're losing carbon, we're losing water, we're losing diversity', a comment which very much chimes with the discourse on rewilding which frames the uplands as 'overgrazed and denuded' and propounds rewilding as a remedy to this (e.g., Carver, 2007). These diametrically opposed ways of viewing the landscape represent what is perhaps the most difficult negotiation for rewilding in Wild Ennerdale, and in England and Britain more broadly. The first perspective values landscape in its current form, as a heritage landscape which, as Max put it, 'reflects how man has interacted with nature over many generations' and therefore resists changes to this. The second derides the current landscape as degraded but values it for its rewilding potential and the scope to restore its ecological processes (Carver, 2007)

While Monbiot's (2013) comment certainly ignited debate about sheep farming in the uplands, it has not been particularly effective as a negotiation technique since it has made many farmers defensive in the face of what they see as attacks on them and their way of life, as Wesley, an environmental farming advisor, explained: 'constantly saying, "it's your fault, it's your fault, it's your fault", going into negotiations saying 'I've come in here to tell you how crap you are', it puts people's backs up'. Philip, a landowner/manager, expressed a similar sentiment, saying that approaches such as this have undermined the 'nuanced conversations' which protagonists in the debate were having and have driven people back to their 'entrenched positions'. Philip explicitly traced the origins of the recent antagonism to *Feral* (George Monbiot's 2013 book) as do scholars who note that *Feral* is 'condemned' for 'polarising and antagonising stakeholders' (Sandom and Wynne-Jones, 2019, p. 222).

Participants did, however, suggest a more constructive way of negotiating the boundary between farming and rewilding, by not viewing the two activities as mutually exclusive. For example, Wesley suggested that successful negotiation was about 'getting people to think it's not all about one thing, it's not all about birds, it's not all about rewilding, it's not all about food production, it's about everything and it's about working with people to put the right thing in the right place'. There was even evidence that this may be beginning to happen, with Brendan, a conservationist, at pains to emphasise that Wild Ennerdale was not seeking to remove sheep farming from the Ennerdale Valley. This distances the Wild Ennerdale Partnership from radical calls, such as those by Monbiot, for sheep farming in the uplands to cease altogether, and attempts to allay the fears of farmers whose view of rewilding has been formed by such comments. Rather, Brendan, and other participants insisted that sheep farming, and farming more generally, could and should continue in the Ennerdale Valley, and that the Wild Ennerdale Partnership recognised the cultural value of this. For example, Brendan stated that Wild Ennerdale was 'certainly not a retreat away from farming a landscape, we recognise [the cultural tradition] and we want to continue that into the future, it's just doing things a bit differently, it's not about ending sheep farming in Ennerdale at all'. Indeed, Brendan saw a place for rewilding and farming to co-exist, albeit with a modified form of farming, for example, with extensive cattle farming as opposed to (relatively more) intensive sheep farming:

> we want that grazing element to continue, but not to as great an extent with sheep, getting hardy cows in that will just bring a different way of grazing within that landscape, and you're still with a farmed animal and it's still an income for the farmer, it's just a different way of farming.

Clare, another conservationist, also saw room for (sustainable) farming and, like Brendan, emphasised the need for better communication to 'promote' the fact that farming and rewilding can co-exist: 'there's a place for sustainable farming, farming with nature, absolutely, and I think we should be promoting that as well, that we can support farming, we can support business, if done well.'

There is a deeper undercurrent in the debate regarding sheep farming in the Lake District, and the uplands more generally, which relates to their status as not being native to Britain. Practitioners stated that sheep are 'not naturally part of our high fells' highlighting that they 'aren't really native here' and noting their Mesopotamian origins. This emphasis on sheep not being native to Britain was linked to the fact that they have, therefore, not co-evolved with British flora, meaning that their mode of grazing is incompatible with sustainable floral growth. It is this incompatible grazing style, they argued, which has led to the 'bare' and 'degraded' condition of the uplands. Based on this argument, there is certainly a move to reduce (although, as per Brendan's comment quoted above, not entirely eliminate) the sheep grazing in Wild Ennerdale. This was acknowledged by practitioners who said that numbers of sheep would be reduced and replaced either with cattle (specifically Black Galloway cattle) or with 'nothing'. Replacing sheep with cattle, participants argued, introduces a different form of grazing, with the cattle performing the role of a large herbivore, acting as a disturbance factor (with their poaching of the ground and their grazing and browsing habits), and filling a trophic niche as an analogue for the extinct aurochs. While the Galloway is a domestic cattle breed and is far smaller than the aurochs would have been, it has the same general form and can, arguably, perform a similar role and function within an ecosystem. Indeed, in contrast to their comments regarding sheep, practitioners emphasised that cattle *were* native to Britain, or at least that their ancestors (the aurochs) were, and were, therefore, much better suited to, and compatible with, the ecosystem of the Ennerdale Valley, the Lake District generally, and Britain more broadly.

As is seen in the wider literature, the debate over the place of native and non-native species is an important one, both in conventional conservation and in rewilding. While conventional conservation is, traditionally, concerned with maintaining a distinction between native and non-native species and with preserving ecosystems in an idealised state *without* non-native species (Katz, 1998; Milton, 2000; Navarro and Pereira, 2015), rewilding is less clear in its stance towards them. It purports to strive to protect environments and ecologies rather than preserving them in a static state which would, in theory, admit the potential which non-native species can offer (Brown, McMorran and Price, 2011; Pearce, 2015; Carver, 2016; Jepson, 2016; Bakker and Svenning, 2018; Wynne-Jones, Strouts and Holmes, 2018). Nonetheless, a reluctance to fully embrace non-native species is evident within the school of rewilding which adopts a 'pristine baseline' approach (i.e., an attempt to return to an idealised ecosystem assemblage free of disturbance from humans or (other) invasive species) (von Essen and Allen, 2016). This was clearly evident in the comments from participants who denigrated sheep as not native and therefore the 'wrong' kind of grazers (Ryder, 1964; Kijas et al., 2012), and yet valorised the role of cattle who *are* seen as native and are, therefore, perceived as interacting differently with Britain's vegetation, playing a valued role as disturbance factors (Orlando, 2015).

Perhaps in light of this, and despite the assurances from practitioners, stakeholders remained concerned about the reduction of sheep numbers in the Ennerdale Valley, maintaining the view that rewilding poses a threat to sheep farming and, in some cases, still considering rewilding as heralding the total removal of farming rather than a shift to other grazing regimes. This was something which Adam, a landowner/manager strongly opposed, seeing it as 'wrong to completely rid it [the Ennerdale Valley] of stock and farming, and a farming community that's been there for thousands of years', repeating the assertion that farming is an intrinsic part of the Lake District: 'commercial activity is part of the Lake District, it's not Yellowstone National Park, it's not an ancient wilderness, it's a working environment'. This highlights one of the key differences between rewilding in the USA versus rewilding in Britain. Rewilding in the USA was very much conceived as involving large areas of wild land with a full suite of species and without human intervention. Such an approach, certainly on so grand a scale, is highly problematic in Britain given the lack of unpopulated areas and the highly anthropogenic nature of landscapes.

Reimagining rewilding

This chapter has discussed the unique landscapes of Wild Ennerdale and the Ennerdale Valley and Lake District within which it is located. The salient landscapes with respect to Wild Ennerdale as a rewilding initiative are the agricultural landscape (especially that of pastoral sheep farming) and the heritage landscape of the Lake District National Park, including the literary legacy of the Romantic poets and the 'idyllic' landscape they evoked which continues to attract national and international visitors. These landscapes are highly valued both for their intrinsic and economic worth, evoking strong sentiments from participants who feel passionately about their individual interpretations of the Ennerdale Valley. The landscapes are valued for the history they contain, the livelihoods they support and the sanctuary they offer. Changes to these landscapes, such as those generated by rewilding, are therefore often regarded critically, with scepticism or even suspicion, and the negotiation of the interface between rewilding and these landscapes is viewed by participants as polarising. Indeed, the term rewilding itself was often received with misapprehension and misgivings, creating confusion and conflict as often as it inspired.

Nonetheless, as is the case in the Avalon Marshes, rewilding is occurring in Wild Ennerdale. Again, rewilding is occurring at a relatively small scale and under a different name – wilding rather than *re*wilding. Human interaction with the landscape continues in the Ennerdale Valley, although it is less interventionist than it has been in the past when the valley was dominated by sheep farming. Instead, extensive cattle farming takes place, with cattle playing a functional role within the socioecosystem and cattle, humans and other species performing socioecolgical processes to co-create a wilder landscape.

Notes

1 A small town four miles west of Ennerdale Bridge.
2 Synonyms and antonyms to 'mess' or 'messy' were also used by participants, for example, 'abandoned' or 'rough' versus 'manicured', 'nice', 'neat', 'tidy', 'uniform' and 'manicured'.

References

Armstrong, P. (2016) *Sheep*. London: Reaktion Books.

Arts, K., Fischer, A. and van der Wal, R. (2012) 'Common stories of reintroduction: A discourse analysis of documents supporting animal reintroductions to Scotland', *Land Use Policy*, 29(4), pp. 911–920.

Arts, K., Fischer, A. and van der Wal, R. (2016) 'Boundaries of the wolf and the wild: A conceptual examination of the relationship between rewilding and animal reintroduction', *Restoration Ecology*, 24(1), pp. 27–34.

Bakker, E. S. and Svenning, J. C. (2018) 'Trophic rewilding: Impact on ecosystems under global change', *Philosophical Transactions of the Royal Society B: Biological Sciences*, 373, pp. 1–6.

Bowker, G. C. (2000) 'Biodiversity data diversity', *Social Studies of Science*, 30(5), pp. 643–683.

Brown, C., McMorran, R. and Price, M. F. (2011) 'Rewilding – A new paradigm for nature conservation in Scotland?', *Scottish Geographical Journal*, 127(4), pp. 288–314.

Carver, S. (2007) 'Rewilding in England and Wales: A review of recent developments, issues, and concerns' in A. Watson, J. Sproull and L. Dean (Ed) *Science and Stewardship to Protect and Sustain Wilderness Values*. Anchorage: U.S. Department of Agriculture, Forest Service, pp. 267–272. Available at: https://www.fs.usda.gov/treesearch/pubs/31040

Carver, S. (2016) 'Rewilding… conservation and conflict', *Ecos*, 37(2), pp. 2–10.

Cassidy, R. (2012) 'Lives with others: Climate change and human-animal relations', *Annual Review of Anthropology*, 41, pp. 21–36.

Ceausu, S. et al. (2015) 'Mapping opportunities and challenges for rewilding in Europe', *Conservation Biology*, 29(4), pp. 1017–1027.

Cloke, P. and Jones, O. (2001) 'Dwelling, place, and landscape: An orchard in Somerset', *Environment and Planning*, 33, pp. 649–666.

Deary, H. and Warren, C. R. (2018) 'Trajectories of rewilding: A taxonomy of wildland management', *Journal of Environmental Planning and Management*, 62(3), pp. 466–491.

DeMello, M. (2012) *Animals and Society: An Introduction to Human-animal Studies*. New York: Columbia University Press.

DeSilvey, C. and Bartolini, N. (2018) 'Where horses run free? Autonomy, temporality and rewilding in the Côa Valley, Portugal', *Transactions of the Institute of British Geographers*, 44, pp. 94–109.

Drenthen, M. (2018) 'Rewilding in layered landscapes as a challenge to place identity', *Environmental Values*, 27(4), pp. 405–425.

Eurostat (2018) *Land Prices Vary Considerably Between and Within Member States*. Luxembourg. Available at: http://ec.europa.eu/eurostat/documents/2995521/8756523/5-21032018-AP-EN.pdf/b1d0ffd3-f75b-40cc-b53f-f22f68d541df

Fairlie, S. (2013) 'Rewilding and food security', *The Land*, 14, pp. 23–25.

Gammon, A. R. (2018) 'The many meanings of rewilding: An introduction and the case for a broad conceptualisation', *Environmental Values*, 27(4), pp. 331–350.

Gray, J. (2014) 'Hefting onto place: Intersecting lives of humans and sheep on Scottish hills landscape', *Anthrozoos*, 27(2), pp. 219–234.

Hayward, M. W. et al. (2019) 'Reintroducing rewilding to restoration – Rejecting the search for novelty', *Biological Conservation*, 233, pp. 255–259.
Head, L. (2015) 'The anthropoceneans', *Geographical Research*, 53(3), pp. 313–320.
Helmreich, S. (2005) 'How scientists think; about "natives", for example. A problem of taxonomy among biologists of alien species in Hawaii', *Journal of the Royal Anthropological Institute*, 11, pp. 107–128.
Höchtl, F., Lehringer, S. and Konold, W. (2005) '"Wilderness": What it means when it becomes a reality – A case study from the southwestern Alps', *Landscape and Urban Planning*, 70(1–2), pp. 85–95.
Hodder, K. H. et al. (2005) *Large Herbivores in the Wildwood and in Modern Naturalistic Grazing Systems*. Peterborough. Available at: publications.naturalengland.org.uk/file/93025
Ingold, T. (2000) *The Perception of the Environment*. London: Routledge.
Jepson, P. (2016) 'A rewilding agenda for Europe: Creating a network of experimental reserves', *Ecography*, 39, pp. 117–124.
Jepson, P. and Blythe, C. (2020) *Rewilding: The Radical New Science of Ecological Recovery*. London: Icon Books.
Jones, P. and Comfort, D. (2020) 'A commentary on rewilding in Europe', *Journal of Public Affairs*, 20(3), pp. 1–6.
Jørgensen, D. (2015) 'Rethinking rewilding', *Geoforum*, 65, pp. 482–488.
Katz, C. (1998) 'Whose nature, whose culture? Private productions of space and the "preservation" of nature' in B. Braun and N. Castree (Eds) *Remaking Reality: Nature at the Millennium*. London: Routledge, pp. 46–63.
Kijas, J. W. et al. (2012) 'Genome-wide analysis of the world's sheep breeds reveals high levels of historic mixture and strong recent selection', *PLoS Biology*, 10(2), pp. 1–14.
Lasanta, T., Nadal-Romero, E. and Arnáez, J. (2015) 'Managing abandoned farmland to control the impact of re-vegetation on the environment. The state of the art in Europe', *Environmental Science and Policy*, 52, pp. 99–109.
Linnell, J. D. C. et al. (2015) 'Framing the relationship between people and nature in the context of European conservation', *Conservation Biology*, 29(4), pp. 978–985.
Lorimer, J. and Driessen, C. (2013) 'Bovine biopolitics and the promise of monsters in the rewilding of Heck cattle', *Geoforum*, 48, pp. 249–259.
Lorimer, J. et al. (2015) 'Rewilding: Science, practice, and politics', *Annual Review of Environment and Resources*, 40, pp. 39–62.
McKibben, B. (1990) *The End of Nature*. London: Penguin.
Milton, K. (2000) 'Ducks out of water. Nature conservation as boundary maintenance' in J. Knight (Ed) *Natural Enemies. People-Wildlife Conflicts in Anthropological Perspective*. London: Routledge, pp. 229–246.
Monbiot, G. (2013) *Feral*. London: Penguin.
Navarro, L. M. and Pereira, H. M. (2012) 'Rewilding abandoned landscapes in Europe', *Ecosystems*, 15(6), pp. 900–912.
Navarro, L. M. and Pereira, H. M. (2015) 'Towards a European policy for rewilding' in L. M. Navarro and H. M. Pereira (Eds) *Rewilding European Landscapes*. London: Springer International Publishing, pp. 205–223.
Nye, D. (2000) 'Technologies of landscape' in D. Nye (Ed) *Technologies of Landscape: From Reaping to Recycling*. Massachusetts: University of Massachusetts Press, pp. 3–17.
Orlando, L. (2015) 'The first aurochs genome reveals the breeding history of British and European cattle', *Genome Biology*, 16(225), pp. 1–3.
Pearce, F. (2015) *The New Wild: Why Invasive Species will be Nature's Salvation*. London: Icon Books.
Pettorelli, N. et al. (2018) 'Making rewilding fit for policy', *Journal of Applied Ecology*, 55 pp. 1114–1125.

Prior, J. and Brady, E. (2017) 'Environmental aesthetics and rewilding', *Environmental Values*, 26(1), pp. 31–51.

Rebanks, J. (2016) *A Shepherd's Life*. London: Penguin.

Rewilding Europe (2020) *Explore our rewilding areas*. Available at: https://rewildingeurope.com/areas/ (Accessed: 21 June 2025).

Root-Bernstein, M., Gooden, J. and Boyes, A. (2018) 'Rewilding in practice: Projects and policy', *Geoforum*, 97, pp. 292–304.

Ryder, M. L. (1964) 'The history of sheep breeds in Britain', *Agricultural History Review*, 12. Available at: http://www.jstor.org/stable/10.2307/40273081

Sandom, C. J. and Macdonald, D. W. (2015) 'What next? Rewilding as a radical future for the British countryside' in D. W. Macdonald and R. E. Feber (Eds) *Wildlife Conservation on Farmland Volume 1: Managing for Nature in Lowland Farms*. Oxford: Oxford University Press, pp. 291–316.

Sandom, C. J. and Wynne-Jones, S. (2019) 'Rewilding a country: Britain as a study case' in N. Pettorelli, S. Durant and J. du Toit (Eds) *Rewilding*. Cambridge: Cambridge University Press, pp. 222–247.

Sandom, C. J. et al. (2013) 'Rewilding', in D. W. Macdonald and K. J. Willis (Eds) *Key Topics in Conservation Biology 2*. Chichester: John Wiley & Sons, pp. 430–451.

Sandom, C. J. et al. (2018) 'Rewilding in the English uplands: Policy and practice', *Journal of Applied Ecology*, 56 pp. 266–273.

Saunders, G. (2016) 'Family ties Human views of Nature -time to see goodness before wildness?', *Ecos*, 37(3–4), pp. 49–54.

Seddon, P. J. et al. (2014) 'Reversing defaunation: Restoring species in a changing world', *Science*, 345(6195), pp. 406–412.

Svenning, J.C. et al. (2016) 'Science for a wilder Anthropocene: Synthesis and future directions for trophic rewilding research', *Proceedings of the National Academy of Sciences of the United States of America*, 113(4), pp. 898–906.

Tanasescu, M. (2017) 'Field notes on the meaning of rewilding', *Ethics, Policy and Environment*. Routledge, 20(3), pp. 333–349.

Thompson, B. L. (1946) *The Lake District and the National Trust*. Kendal: Titus Wilson and Son.

Tree, I. (2018) 'Creating a mess -The Knepp rewilding project', *Bulletin of the Chartered Institute of Ecology and Environmental Management*, 100, pp. 29–34.

Tsing, A. (2012) 'Unruly edges: Mushrooms as companion species', *Environmental Humanities*, 1, pp. 141–154.

United Utilities (2020) *Ehen Compensatory Measures*. Warrington: United Utilities.

van der Zanden, E. H. et al. (2017) 'Trade-offs of European agricultural abandonment', *Land Use Policy*, 62, pp. 290–301.

Visit Cumbria (2020) *Wastwater*. Available at: https://www.visitcumbria.com/wc/wastwater/ (Accessed: 21 June 2025).

von Essen, E. and Allen, M. (2016) 'Wild, but not too-wild animals: Challenging goldilocks standards in rewilding', *Between the Species*, 19(1), pp. 80–108.

White, R. (1995) '"Are you an environmentalist or do you work for a living?": Work and nature' in W. Cronon (Ed) *Uncommon Ground: Rethinking the Human Place in Nature*. New York: W.W. Norton and Co, pp. 171–185.

Wild Ennerdale (2006) *Wild Ennerdale Stewardship Plan*. Available at: http://www.wildennerdale.co.uk/ (Accessed: 21 June 2025).

Wild Ennerdale (2019) *Wild Ennerdale*. Available at: http://www.wildennerdale.co.uk/ (Accessed: 21 June 2025).

Wynne-Jones, S., Strouts, G. and Holmes, G. (2018) 'Abandoning or reimagining a cultural heartland? Understanding and responding to rewilding conflicts in Wales - The case of the Cambrian Wildwood', *Environmental Values*, 27(4) pp. 377–403.

4 Domesticating rewilding in England and Britain

Adapting the factors that can confer family resemblance to rewilding

So far, this book has introduced rewilding and the factors that can confer family resemblance to it (Chapter 1), discussed how rewilding relates to England specifically and Britain more generally (Chapter 1), and examined the Avalon Marshes and Wild Ennerdale as examples of rewilding in England (Chapters 2 and 3, respectively). Having done so, it is possible to say more about rewilding in England and by extension Britain. Commonalities are evident in the rewilding of the Avalon Marshes and Wild Ennerdale, including significant variations from the way the concept is applied in other parts of the world. That these commonalities exist, despite the different contexts of the Avalon Marshes and Wild Ennerdale, makes it possible to extrapolate from them to draw conclusions regarding rewilding in England and Britain more widely. A unique form of rewilding was revealed and this chapter discusses the notion that rewilding in England and Britain is being 'domesticated': it is being adapted to function alongside and along with humans in order to coexist with other land uses.

There are two important points to note here. First, this chapter attempts to report the finding that rewilding is being domesticated rather than to advocate for it. Secondly, it should be acknowledged that, notwithstanding the first point, this chapter does present the domestication of rewilding as a 'win-win' scenario. Mutual compromise by rewilding and other land uses (such as farming) are presented as offering a 'best-of-both-worlds' scenario in which rewilding can progress (where it might not otherwise be able to do so), contributing to ecological gains, while human activity can also progress (where it might otherwise have to cease), contributing to social, cultural and economic gains. Another framing, likely to be adopted by rewilding purists, is that this is a 'lose-lose' or 'worst-of-both-worlds' scenario. Going further, they might argue that what is presented here has ceased to be rewilding at all. Rewilding in England, and Britain, is certainly very different from rewilding in other parts of the world, but this does not preclude it from being considered rewilding. Rewilding's plasticity allows a spectrum of activity to be considered

DOI: 10.4324/9781003456490-5

rewilding. According to the typology proposed in Chapter 1, the Avalon Marshes and Wild Ennerdale are examples of 'culture led, active nature' and 'nature led, active culture' rewilding, respectively. More broadly, they both display factors which afford them family resemblance to rewilding, albeit in a modified form. The modification, or *domestication*, of these factors is the focus of this chapter.

First, the term rewilding itself is examined, being modified to become 'wilding' (or perhaps wild, wilds or wilder). Abandoning the prefix 're' helps to avoid any concerns about its regressive implications and also disassociated projects from wider debates about rewilding. Second, the scale of rewilding is reduced – rewilding in England, and Britain, not only operates at a smaller scale than in other parts of the world but also at a smaller scale than it might aspire to. Third, while biodiversity is increased, it is not increased to the extent that the full suite of species are present. Fourthly, and relatedly, full ecological function is not restored. Fifthly, partly, although by no means exclusively, as a result of this, human intervention in the form of land and species management continues, albeit it at a reduced and/or in a modified form. Lastly, and again as a result of other modifications, particularly on-going human intervention, other-than-human agency is not fully restored.

Changing terminology: from rewilding to wilding

Neither the Avalon Marshes nor Wild Ennerdale use the term rewilding in their names. While both projects were established prior to the rise of rewilding's popularity as a term it is still notable that neither site has elected to alter their name to adopt it (as will be seen this is something which is equally marked in other British rewilding sites). The Avalon Marshes avoid the term rewilding altogether, although it is applied to them by others, while Wild Ennerdale holds it at arm's length, associating with it indirectly through the European Rewilding Network and through the use of 'wild' in its appellation. Conservation projects distancing themselves from the term rewilding is indicative of the major questions regarding its use in England, and Britain more widely. Perhaps the most striking examples of this come from Wales where Cambrian Wildwood/Coetir Anian and Summit to Sea/O'r Mynydd i'r Môr both reduced their association with rewilding; Cambrian Wildwood/Coetir Anian abandoning its use of the term (Wynne-Jones, Strouts and Holmes, 2018; Coetir Anian/Cambrian Wildwood, 2020) and Summit to Sea/O'r Mynydd i'r Môr dissolving its relationship with Rewilding Britain (Forgrave, 2019).

In place of 'rewilding', projects are adopting other terms such as wild, wilds, wilder or wilding. Wild Ennerdale is of course a good example of this but it is by no means alone. Since the start of the research on which this book is based, the number of 'rewilding' projects listed on the Rewilding Britain and Rewilding Europe websites has proliferated. While projects are embracing rewilding by associating themselves with Rewilding Britain and Rewilding Europe, they are also distancing themselves from it by not calling themselves rewilding.

For example, at time of writing, there are nine English and nine Scottish 'rewilding' projects on the Rewilding Europe website. Of the English projects, while six use the word 'wild' or 'wilder' none use the term rewilding. Of the Scottish projects, one uses 'wild', another uses 'wilding' and two use the term rewilding. Thus, out of 18 projects only two explicitly identify as rewilding, both of which are Scottish rather than English. On Rewilding Britain's website, there are 55 English, 14 Scottish and 4 Welsh 'rewilding' projects. Of the English projects, 25 use 'wild', 'wilds', 'wilder' or even 'wilding' but only one uses the term 'rewilding' (although it should be noted that this is an urban rather than a rural project). Of the Scottish projects, one uses the term 'wild' and two use 'rewilding'. Of the Welsh projects, only one uses a term related to wildness, choosing the term 'wilder'. Here, 3 from a possible 73 projects choose to title themselves rewilding, 2 of which are in Scotland and only 1 in England. This being said, there may be other projects in Britain that call themselves rewilding but are not listed on the Rewilding Britain or Rewilding Europe websites. It is also possible that projects are listed on these websites by one name, which does not include 'rewilding', but also have another name which does. For example, a project listed on Rewilding Britain's website as 'Upcott Grange Farm' calls itself 'Rewilding Coombeshead' on its own website. It should be noted here that this project was founded by Derek Gow; Gow is discussed in the final chapter of this book as a 'maverick rewilder' and, as will be seen, is not one to shy away from the controversy associated with the term rewilding so could be considered something of an anomaly here.

Other projects, particularly English projects, are clearly avoiding the use of the term 'rewilding', choosing wild, wilds, wilder or wilding, all without the prefix 're'. By doing so, these sites are not overtly identifying themselves as rewilding projects, thus avoiding the general controversy associated with it. Moreover, they are disassociating themselves from one of the specific criticisms levelled at rewilding in that the term has implications of being regressive or backward looking (Sandom, Hughes and Macdonald, 2013; Jørgensen, 2015; Carver, 2016; Deary and Warren, 2018). Abandoning the prefix 're' avoids this implication with 'wilding' having connotations of making land wilder without the implications of returning or restoring it to a previous state. This suggests that terms such as 'wild' or 'wilding' (as opposed to 'rewild' or 'rewilding') may be more acceptable in the English and British contexts since wilding simply has connotations of being any degree wilder than the current state while still permitting management (Prior and Brady, 2017; van Horn, 2017). *R*ewilding by contrast not only has connotations of returning to a previous, wild state (Sandom, Hughes and Macdonald, 2013; Jørgensen, 2015; Carver, 2016; Deary and Warren, 2018) but also has connotations of increasing other-than-human agency to the extent that little or no human intervention is possible or permitted (Höchtl, Lehringer and Konold, 2005; Navarro and Pereira, 2012; Lorimer et al., 2015; Tanasescu, 2017; Gammon, 2018; Pettorelli et al., 2018) with neither of these being palatable or even plausible in England and Britain.

As an illustration, 'Wilder Blean' is a 'wilding' project in Kent in the southeast of England whose webpage describes wilding as occurring 'when nature is given the tools and space it needs to recover itself' (Kent Wildlife Trust, 2020), a definition which is very difficult to distinguish from that of rewilding. Indeed, in a newspaper article, the project manager was quoted as situating Wilder Blean within the rewilding debate: 'sometimes in the rewilding debate people think that it's a look back to the past, but that's not what we're about. We're about trying to find the right natural solution for the modern world' (Stan Smith, Wilder Landscapes Manager at Kent Wildlife Trust and leader of the Wilder Blean Project quoted in Carrington (2020)). Wilder Blean has apparently adopted the rewilding philosophy but has chosen wilding as a more 'acceptable' term. If Wilder Blean is indicative of other projects, this would suggest that while rewilding is not being used in the naming of projects it is being employed in describing and discussing them. Thus, rewilding is being used when more in-depth and nuanced conversations can be had, so that practitioners can explain what *they* mean by rewilding and what it means in the context of their particular project.

Changing scale: from large scale to 'large scale'

Rewilding at the Avalon Marshes covers an area of 1500 hectares and Wild Ennerdale extends to 4300 hectares. Neither of these project areas would be considered large by European standards where rewilding projects such as Swedish Lapland can be over three million hectares in size (Rewilding Europe, 2020a). Even by British rewilding standards, or perhaps more accurately aspirations, these projects are not large. Alastair Driver, former director of Rewilding Britain (which describes rewilding as the 'large-scale restoration of ecosystems'), defined 'large scale' as '10,000 ha plus in England, Wales and Northern Ireland and 100,000 ha plus in Scotland' (Catchments, 2018). The Avalon Marshes and Wild Ennerdale are clearly far smaller than this and, as a point of comparison, the other English rewilding projects in the European Rewilding Network are also well below 10,000 hectares: Wilder Blean – 560 ha, Wild Ken Hill – 425 ha, Wallasea Island Wild Coast Project – 850 ha, Knepp Wildland – 1400 ha, and Wicken Fen Vision – 5300 ha (Rewilding Europe, 2020b).

'Large scale' then is, arguably, a relative measure, meaning something different in England than it does in continental Europe or even within Britain, especially since, in his definition, Driver makes a distinction between the scale of rewilding in England (and also Wales and Northern Ireland) and Scotland. It could be argued that rewilding in England does operate at 'large-scale', but that the scale is 'large' only within the English context and is small by comparison to the scope of rewilding projects in other countries. This suggests that rewilding in England has developed on a much smaller scale than might be considered in countries with larger and more sparsely populated land areas, as a way of tailoring it to the English context. A similar argument can be made about Scotland since the Scottish rewilding projects in the European Rewilding Network are also well below the 100,000 hectares which Driver advocated: Bamff Wildland – 526 ha, Coigach Assynt – 60000 ha, Highlands Rewilding – 862 ha, Rewilding the Scottish Highlands – 12000 ha, Saving Wildcats – 59000 ha, Uist Forest Retreat – 560 ha (Rewilding Europe, 2020b).

Changing form: from biodiversity to functional diversity

Partly, although not exclusively, because of the scale at which rewilding operates in England and Britain, it is not practicable to increase biodiversity to the point which the IUCN (2021) rewilding principles suggest – that rewilding should 'aim to restore the complete or near-complete food-web'. While, to a considerable extent, the barriers to restoring Britain's complete or near-complete food web are physical (linked to the size of available habitat for species reintroductions) there are also considerable socio-cultural barriers, especially in relation to other IUCN (2021) principles which emphasise the role of 'large herbivores and apex predators' in 'maintaining and enhancing' biodiversity and restoring trophic interactions. Publics can be reluctant to accept *any* species reintroductions, especially when, as will be seen in Chapters 6 and 7, there is the potential for human interests or values to be affected. This reluctance is exacerbated in the case of apex predators and is particularly pronounced in Britain since all apex predators were extirpated centuries ago (see, e.g., Neilson, 2019).

Apex predators were, therefore, conspicuous by their absence from the Avalon Marshes and Wild Ennerdale and neither site has plans to reintroduce any. Indeed, research participants were at pains to *de*emphasise the role of apex predator reintroduction in rewilding projects in England. This is a marked contrast to rewilding agendas in other locations which focus heavily on apex predator reintroductions (Soule and Noss, 1998; Deary and Warren, 2018). Interestingly, the narrative in Scotland is subtly different with a greater focus on the reintroduction of apex predators, possibly related to the lower population density (England has a population density of 432 people/square kilometre, while Scotland has a population density of 70 people/square kilometre (ONS, 2020)) and the, relatively, larger scale at which rewilding can and does operate there (see Warren, 2002; Hetherington, 2005; Arts, Fischer and van der Wal, 2012).

There are several important points to note in relation to the absence of apex predators from the rewilding agenda in England and, albeit to a lesser extent, Britain as a whole. First, the absence of apex predators means that not all trophic interactions are restored. As a result, humans often intervene, replacing other-than-human processes with human interaction. Secondly, the absence of apex predators allows for a relatively permissive mode of biopolitics in relation to species as self-determining agents – species are 'let live, let die'. Were apex predators to be introduced it is highly likely that this mode of biopolitics would change to become considerably less permissive. We can see a foreshadowing of this in Chapters 6 and 7 which discuss the introduction of small predators to Britain (the red kite (*Milvus milvus*) and the European wildcat (*Felis silvestris*). Considerable controversy has surrounded the reintroduction of these two species, despite the fact that they do not pose a threat to humans, and there are already calls for the logics of death as applied to red kites to be reviewed (see Chapter 6). We can extrapolate from this to conclude that, were apex predators to be introduced, it is likely that there would be calls for a logic of make die rather than let die in relation to them.

Aside from apex predators, the IUCN (2021) rewilding principles mention large herbivores. Large herbivores are certainly being enrolled in rewilding projects in England and Britain both to enhance biodiversity in and of themselves and also through their trophic interactions. It should be noted, however, that in many cases these large herbivores are domestic analogues of their wild counterparts and could therefore perhaps be said to contribute *functional diversity* as much as biodiversity (see Chapter 5). Indeed, the IUCN (2021) recognises the value of such analogues where they have 'similar ecological and trophic functionality' as the species they are replacing.

Changing perspective: from apart from to a part of

The factors which can confer family resemblance to rewilding are densely entangled – the scale at which a project operates can affect which species can be involved which in turn affects which ecological processes are present, this in turn can affect human decision making regarding how much or how little to intervene which in turn affects how much or how little humans control other-than-human agency. Thus, rewilding's ambition to increase ecological processes is highly contingent on the scale at which it operates and the species, and, indeed, the abiotic elements, involved. Ecological functioning is also contingent on the extent to which the other-than-human species and elements involved in rewilding can exert their (wild) agency or the extent to which this is controlled by humans.

Because rewilding in England and Britain tends to occur at smaller scales than rewilding in other parts of the world, and because rewilding in England and Britain is unable to restore the complete suite of species associated with its ecosystems, some ecological processes are, unavoidably, missing. The domesticated form of rewilding present in England and Britan has not diverted from the fundamental tenet of rewilding which aims to restore *all* ecological processes to an ecosystem, rather it has recognised that this is not immediately achievable and, indeed, may never be. Domesticated rewilding has reconciled itself to this and, rather than taking a purist view that rewilding is fatally compromised by its inability to restore all ecological function, it accepts the constraints within which it is working and increases what ecological processes it can. Where it is not possible for ecological functioning to be restored, humans often step in to replace these missing processes. Following this logic to its inevitable conclusion, rewilding purists would (correctly) point out that compromising on one of rewilding's tenets leads to further compromises in regard to its other tenets. Domesticated rewilding accepts and even embraces these compromises, embracing the opportunity for humans to play an active role in rewilding and in socioecological systems.

Changing approach: from intervention to interaction

The roles that humans can play in rewilding are diverse. To a large extent, they involve performing roles that are not being filled by other members of the ecosystem, often because these members are absent (e.g., humans cull deer in Scotland which might otherwise be hunted by wolves). It is also fair to say that humans are most likely to step in to perform roles where not doing so might affect human values or interests (e.g., humans often intercede in waterways and water flow if homes are likely to be flooded). In England and Britain such intervention is much more acceptable than it is in rewilding in other places, particularly the USA where rewilding, by definition, excludes humans (e.g., Rewilding Earth, 2020). Thus, while the overall aim of rewilding in England and Britain remains the reduction of human intervention, it permits considerably more intervention than is typical of other rewilding in other areas. This is largely a pragmatic response to the fact that rewilding in England and Britain needs to be able to co-exist with humans and other forms of land use. It may also represent a shift in approach from seeing any human involvement in rewilding as *intervention*, and therefore negative, to viewing it as *interaction*, and therefore not only acceptable but even positive. One of the most common ways this occurs in practice is through pastoral agriculture (although other forms of other-than-human animal management also represent significant ways that humans intercede in rewilding).

Rewilding's interface with farming is highly significant in the Avalon Marshes and Wild Ennerdale, perhaps particularly in Wild Ennerdale where the farming tradition not only provides livelihoods but also a way of life which is recognised as having heritage and cultural value (UNESCO, 2017). Extensive agriculture is continuing at both sites as part of the rewilding projects: while ideologically very different, the goals of rewilding and extensive farming are not necessarily incompatible and, through mutual compromise, can lead to mutual benefits (Mondière et al., 2022; Thomas et al., 2022) For example, changes in farming approaches can have significant benefits to biodiversity and ecological functioning and, rather than the assumed decrease, can also result in increased agricultural productivity (Jackson, Maginnis and Sengupta, 2007; Neely and Hatfield, 2007; Thompson et al., 2007). Equally, changes in rewilding approach, such as the use of traditional livestock breeds rather than non-native, non-domestic species to serve as analogues and proxies, can have the same (or at least similar) ecological impact while still allowing productive agriculture to continue. While such approaches contribute to rewilding's domestication, their adoption can allow farming and rewilding to cooperate to their mutual benefit and foster collaboration towards compatible rather than competing interests.

By allowing human interaction with rewilding, based on extensive farming models such as at the Avalon Marshes and Wild Ennerdale, rewilding in England and Britain is developing in a way which is tailored to its cultural contexts. The adaptation makes it feasible to maintain the sociocultural value of rural communities, traditional farming practices and heritage landscapes rather than attempting to remove or erase signs of human intervention in line with other interpretations of rewilding (Jørgensen, 2015; Deary and Warren, 2018).

Changing agency: from human to other-than-human[1]

As a result of English and British rewilding's alignment with extensive agriculture, domestic species, particularly cattle, are commonly enrolled in rewilding projects (as will be discussed more fully in Chapter 5). These cattle are performing naturalistic grazing and, in doing so, are serving as analogues for the large wild herbivores which have been extirpated from Britain. They also serve as an exemplar of the new modes of biopolitics emerging in England and Britain as a result of rewilding. These new biopolitical modes reflect new human–animal relationships and the changing extent to which other-than-human animals can exert their agency as a result of the restoration of control from human to other-than-human. It should be noted, however, that as with many of the modifications made in English and British rewilding to the factors conferring family resemblance to rewilding, the increase in other-than-human agency has been one of degree rather than one of kind. Rewilding in England, and Britain, aspires to increase other-than-human agency and, indeed, does so, but other-than-human agency remains more constrained in English and British rewilding projects than in rewilding projects in other parts of the world. This is evident from the fact that while many of the new biopolitical modes in operation in the Avalon Marshes and Wild Ennerdale, and by extension England and Britain, are *ideologically* different from those which operate in farming or conventional conservation, in *practice* they operate in very similar ways. A considerable amount of control is still exerted over other-than-human species (and, indeed, over the other-than-human world more generally) with these biopolitical modes still relatively closely aligned to the classic Foucauldian 'make live and let die' logic (Foucault, 1976) rather than demonstrating the radical tolerance for other-than-human agency or the paradigm shift in human–nature relations which rewilding both calls for and aspires to (Campbell, 2006; IUCN, 2021).

Domesticating rewilding

The modifications to the factors that can confer family resemblance to rewilding, discussed above, illustrate the unique nature of rewilding in England, and Britain, which is distinct from rewilding in other locations. This is largely due to the way that rewilding is re-shaping itself in the English and British context.

Rewilding's local tailoring to accommodate and adapt to the English and British contexts is domesticating it – some of its more radical potential is being moderated, making it more compatible with existing land use and making it easier for rewilding to operate with and alongside other land uses. What emerges is a uniquely English, and British, form of rewilding which incorporate a range of practices which seek to increase biodiversity and restore ecological functioning but which do so with greater or lesser levels of human intervention and, by extension, greater or lesser opportunity for the expression of other-than-human agency. Other-than-human processes are encouraged or compelled, while other-than-human agency is permitted or controlled.

The uniquely English and British expressions of rewilding illustrate how the concept of rewilding has travelled from North America, where it originated, and been stretched (Collier and Mahon, 1993) from its original meaning given its plasticity (Star and Griesemer, 1989; Poerksen, 1995; Bensaude-Vincent, 2014; Jørgensen, 2015). In light of all the local tailoring (Star and Griesemer, 1989) evident in relation to rewilding in the English and British context, the factors outlined in Chapter 1 as being able to confer family resemblance to rewilding, have been modified to reflect this unique expression of rewilding (see Table 4.1). Despite being scaled down in terms of size, and, indeed,

Table 4.1 Adaptations to the factors deemed capable of conferring family resemblance to rewilding as they apply to rewilding in the English, and British, contexts

Factor	*Adaptation*
Identifying as rewilding	Rewilding is increasingly self-identifying as 'wilding' rather than 'rewilding' in project titles although rewilding will often appear in project descriptions and its concepts are often applied in practice.
Functioning at large scale	Rewilding operates at a smaller scale than it does in continental Europe and on the continents of North and South America.
Increasing biodiversity	Rewilding aims to increase biodiversity but does so within certain limits, most notably without reintroducing mega-fauna species, particularly apex predators but also large herbivores and omnivores.
Increasing ecological functioning	Increasing ecological functioning is a goal of rewilding but is, to an extent, restricted by the inability to reintroduce certain species: as a result not all trophic niches are filled which in turn affects trophic cascades and other ecosystem functions.
Reducing human intervention	Reducing human intervention is an ambition of rewilding but only up to a certain point, it being neither practicable nor acceptable to completely withdraw human management of landscapes (due in part to the way other rewilding factors, namely increase in biodiversity and ecological functioning, are compromised).
Increasing other-than-human agency	The increase of other-than-human agency is an aspiration of rewilding, but this is limited by the requirement for some level of human intervention which, by definition, indeed, by design, impinges on other-than-human agency.

some of its more radical ideals, and despite not necessarily always using the term rewilding explicitly, rewilding is still sufficiently novel and inspiring to be able to offer something new and innovative to conservation in England and Britain.

Note

1 This section, and indeed this book, focusses on other-than-human animals in providing examples of the points made. It should be noted, however, that this is not to suggest that the points made are not equally applicable to plants or indeed the abiotic elements involved in rewilding.

References

Arts, K., Fischer, A. and van der Wal, R. (2012) 'Common stories of reintroduction: A discourse analysis of documents supporting animal reintroductions to Scotland', *Land Use Policy*, 29(4), pp. 911–920.

Bensaude-Vincent, B. (2014) 'The politics of buzzwords at the interface of technoscience, market and society: The case of "public engagement in science"', *Public Understanding of Science*, 23(3), pp. 238–253.

Campbell, T. (2006) '"Bios," immunity, life: The thought of Roberto Esposito', *Diacritics*, 36(2), pp. 2–22.

Carrington, D. (2020) 'Wild bison to return to UK for first time in 6,000 years', *The Guardian*. Available at: https://www.theguardian.com/environment/2020/jul/10/wild-bison-to-return-to-uk-kent

Carver, S. (2016) 'Rewilding… conservation and conflict', *Ecos*, 37(2), pp. 2–10.

Catchments (2018) *Time to rewild?* Available at: https://www.catchments.ie/time-to-rewild-a-view-from-across-the-irish-sea/ (Accessed: 21 June 2025).

Coetir Anian/Cambrian Wildwood (2020) *Land Management, Coetir Anian/Cambrian Wildwood*. Available at: https://www.cambrianwildwood.org/land-management (Accessed: 21 June 2025).

Collier, D. and Mahon, J. E. (1993) 'Conceptual "Stretching" revisited: Adapting categories in comparative analysis', *The American Political Science Review*, 87(4), pp. 845–855.

Deary, H. and Warren, C. R. (2018) 'Trajectories of rewilding: A taxonomy of wildland management', *Journal of Environmental Planning and Management*, 62(3), pp. 466–491.

Forgrave, A. (2019) 'Farming anger forces Rewilding Britain to pull out of Summit to Sea project in Mid Wales', *Daily Post*, 21 October. Available at: https://www.dailypost.co.uk/news/local-news/rewilding-britain-summit-sea-wales-17119541

Foucault, M. (1976) *'Society Must Be Defended' Lectures at the College de France, 1975–76*. New York: Picador.

Gammon, A. R. (2018) 'The many meanings of rewilding: An introduction and the case for a broad conceptualisation', *Environmental Values*, 27(4), pp. 331–350.

Hetherington, D. (2005) *The Feasibility of Reintroducing the Eurasian Lynx (Lynx lynx) to Scotland*. Aberdeen: University of Aberdeen.

Höchtl, F., Lehringer, S. and Konold, W. (2005) '"Wilderness": What it means when it becomes a reality - A case study from the southwestern Alps', *Landscape and Urban Planning*, 70(1–2), pp. 85–95.

IUCN CEM (2021) *Rewilding Principles*.Gland: IUCN. Available at: https://iucn.org/sites/default/files/2022-10/principles_of_rewilding_cem_rtg.pdf

Jackson, W., Maginnis, S. and Sengupta, S. (2007) 'Planning at landscape scale' in S. Scherr and J. McNeely (Eds) *Farming with Nature: The Science and Practice of Ecoagriculture*. Washington: Island Press, pp. 308–321.
Jørgensen, D. (2015) 'Rethinking rewilding', *Geoforum*, 65, pp. 482–488.
Kent Wildlife Trust (2020) *Wilder Blean*. Available at: https://www.kentwildlifetrust.org.uk/wilderblean (Accessed: 21 June 2025).
Lorimer, J. et al. (2015) 'Rewilding: Science, practice, and politics', *Annual Review of Environment and Resources*, 40, pp. 39–62.
Mondière, A., Thomas, V., Corson, M.S., Diraison, M., Dulac, P., Siguier, D. and van der Werf, H. (2022) 'Farming with nature: lessons from rewilding agriculture and Paysans de nature' in D. Bruce and A. Bruce (Eds), *Transforming Food Systems: Ethics, Innovation and Responsibility*, Wageningen: Wageningen Academic, pp. 145–151.
Navarro, L. M. and Pereira, H. M. (2012) 'Rewilding abandoned landscapes in Europe', *Ecosystems*, 15(6), pp. 900–912.
Neely, C. and Hatfield, R. (2007) 'Livestock systems' in S. Scherr and J. McNeely (Eds) *Farming with Nature: The Science and Practice of Ecoagriculture*. Washington: Island Press, pp. 121–142.
Neilson, A. (2019) 'Disenchanted natures: A critical analysis of the contested plan to reintroduce the Eurasian Lynx into the Lake District National Park', *Capitalism, Nature, Socialism*, 32(1), pp. 107–125.
ONS (2020) *Population estimates for the UK, England and Wales, Scotland and Northern Ireland, provisional: mid-2019*. Available at: https://www.ons.gov.uk/peoplepopulationandcommunity/populationandmigration/populationestimates/bulletins/annualmidyearpopulationestimates/mid2019 (Accessed: 21 June 2025).
Pettorelli, N. et al. (2018) 'Making rewilding fit for policy', *Journal of Applied Ecology*, 55, pp. 1114–1125.
Poerksen, U. (1995) *Plastic Words: The Tyranny of a Modular Language*. University Park: Pennsylvania State University Press.
Prior, J. and Brady, E. (2017) 'Environmental aesthetics and rewilding', *Environmental Values*, 26(1), pp. 31–51.
Rewilding Earth (2020) *What is rewilding*. Available at: https://rewilding.org/what-is-rewilding/ (Accessed: 21 June 2025).
Rewilding Europe (2020a) *Swedish Lapland*. Available at: https://rewildingeurope.com/rew-project/lapland-rewilding-area/ (Accessed: 21 June 2025).
Rewilding Europe (2020b) *European Rewilding Network*. Available at: https://rewildingeurope.com/european-rewilding-network/ (Accessed: 21 June 2025).
Sandom, C. J., Hughes, J. and Macdonald, D. W. (2013) 'Rooting for rewilding: Quantifying wild boar's sus scrofa rooting rate in the Scottish Highlands', *Restoration Ecology*, 21(3), pp. 329–335.
Soule, M. and Noss, R. (1998) 'Rewilding and biodiversity: Complementary goals for continental conservation', *Wild Earth*, 8(3), pp. 18–28.
Star, S. L. and Griesemer, J. R. (1989) 'Institutional ecology, "translations" and boundary objects: Amateurs and professionals in Berkeley's museum of vertebrate zoology, 1907–39', *Social Studies of Science*, 19, pp. 387–420.
Tanasescu, M. (2017) 'Field notes on the meaning of rewilding', *Ethics, Policy and Environment*, 20(3), pp. 333–349.
Thomas, V., Mondière, A., Corson, M.S., and van der Werf, H. (2022) 'Domesticating rewilding: combining rewilding and agriculture offers environmental and human benefits', in D. Bruce and A. Bruce (Eds), *Transforming Food Systems: Ethics, Innovation and Responsibility*, Wageningen: Wageningen Academic, pp. 165–170.
Thompson, J. et al. (2007) 'Biodiversity in agroecosystems' in S. Scherr and J. McNeely (Eds) *Farming with Nature: The Science and Practice of Ecoagriculture*. Washington: Island Press, pp. 46–60.

UNESCO (2017) *The English Lake District*. Available at: https://whc.unesco.org/en/list/422/ (Accessed: 21 June 2025).

van Horn, G. (2017) 'Into the wildness' in G. van Horn and J. Hausdoerffer (Eds) *Wildness: Relations of People and Place*. Chicago: University of Chicago, pp. 1–8.

Warren, C. (2002) *Managing Scotland's Environment*. Edinburgh: Edinburgh University Press.

Wynne-Jones, S., Strouts, G. and Holmes, G. (2018) 'Abandoning or Reimagining a Cultural Heartland? Understanding and Responding to Rewilding Conflicts in Wales - the case of the Cambrian Wildwood', *Environmental Values*, 27(4), pp. 377–403.

5 Let live and let die

The biopolitics of rewilding and (self)-governing cows

Rewilding is not a solely human endeavour, deliberately or otherwise, and more or less consciously, it involves the other-than-human world. Despite rewilding's oft-stated intention to reduce human intervention in 'natural' processes, the relationship between the human and other-than-human agents involved is often an unequal one, with humans frequently unwilling or unable to (entirely) relinquish control, asserting biopolitical control over the other-than-human animals involved rather than respecting their agency.

Of the other-than-human animals involved in rewilding in Britain, domestic cattle (*Bos taurus*) are one of the most commonly seen: indeed, in some respects cattle are the rewilding 'tool of choice' for rewilding projects. As the term 'tool' implies, there can be a utilitarian element to the way cattle are viewed and treated in relation to rewilding and this places them in an interesting position: while they may be valued for the role they play as a rewilding tool, simultaneously, their agency in this process may not necessarily be valued or respected. Furthermore, as a domestic species, even when cattle are involved in rewilding projects, their lives are regulated by humans, meaning that the cattle occupy a liminal state between wildness and domestication.

New biopolitical modes have been established to govern the lives and deaths of the cattle involved in rewilding and to negotiate the frictions and compromises which the new forms of human–cattle relationships inherent in rewilding reveal. This chapter explores these relationships and biopolitical modes as they apply to cattle involved in rewilding at the Avalon Marshes and Wild Ennerdale. Four new biopolitical modes, specific to rewilding are identified: 'expendable objects', 'animal machines/human proxies', 'analogues for other species' and 'self-determining agents'. Cattle in each of these modes have different degrees of agency and are subject to different levels of human intervention. Furthermore, their lives and deaths, as individuals and as species, are considered quite differently by those involved in rewilding and who are, ultimately, responsible for them.

DOI: 10.4324/9781003456490-6

The wild/domestic cattle and other animals involved in rewilding in Britain

Britain's largest *extant*, native, herbivore is the red deer (*Cervus elaphus*). Meanwhile, Britain's *extinct* large, native herbivores include aurochs (*Bos primigenius*), elk (*Alces alces*) and wild horses (*Equus ferus*). As a result of these extinctions, the role that these species would have played in Britain's ecosystem is missing. On the other hand, these roles are filled, up to a point, by domestic cattle (*Bos taurus*) and domestic horses (*Equus ferus caballus*). Indeed while, as a global movement, rewilding is advocating the reintroduction of missing species, rewilding in Britain is taking a pragmatic approach and replacing missing species with domestic 'analogues' – species which, while different from the missing species, can perform a similar function.[1] Domestic cattle, in various guises, are in many ways, the 'tool of choice'[2] for rewilding programmes looking for a large herbivore to fulfil the grazing and browsing functions once performed by aurochs, elk and wild horses. The reasons for this are manifold and it is logical to consider the other candidate species, specifically deer and horses, before going on to discuss domestic cattle in more detail. Elk are not currently present in Britain outside captivity and therefore do not feature prominently in the rewilding discourse so are not discussed here. Aurochs, or aurochs analogues, are discussed alongside domestic cattle.

Red deer are, as mentioned, a wild species still extant in Britain. As a wild species, enrolling red deer in rewilding projects would, therefore, perhaps be more aligned to rewilding's ethos of increasing biodiversity and increasing other-than-human agency than enrolling domestic species. The human–deer relationship is, however, complex and like cattle deer are, in many ways, a liminal animal; deer exist at the threshold of wildness and domestication, simultaneously *both* wild *and* domestic, and *neither* wild *nor* domestic (Wischermann and Howell, 2018). This status of deer has been discussed eloquently by scholars who describe how the overlap between human intervention in the lives of 'wild' deer and the hands-off 'management' of 'domestic' deer means that it can be 'difficult to say under which contexts animals are wild and under which they are domestic' (Linnell et al., 2015, p. 981). Furthermore, intervention in the lives of wild deer and management of domestic deer is contested, with debates over how deer (wild or domestic) should, or should not, be fed, fenced or otherwise controlled (Linnell et al., 2015). The inclusion of red deer (or indeed any of the other species of deer present in Britain) in rewilding projects would, therefore, not be straightforward and may, or may not, require the use of feeding, fencing and culling or slaughter which rewilding practitioners, proponents and stakeholders may see as antithetical to rewilding's ethos of non-intervention.[3]

The human–horse relationship is also complicated in Britain. This is partly because, like deer, some horses are in a liminal state between wild and domestic. While there are no longer any wild horses in Britain there are free-living horses which are members of the domestic species and which are often referred to as 'feral' or 'semi-wild'. It is these horses which exist in a liminal state – owned and, to a greater or lesser extent, managed by humans but living largely independently of humans and therefore exerting their own agency. Scholars have suggested that the terms '"wild" and "domestic" do not adequately describe the characteristics of the ongoing relationship between people and horses' (DeSilvey and Bartolini, 2018, p. 9), with people retaining a strong sense of responsibility for and affinity with horses even when the horses are living in a 'wild' or 'natural' state.

Part of this is because, in Britain, horses are viewed quite differently from cattle or deer, they are seen solely as companion animals and not considered a food animal: the idea of consuming horse meat can evoke something akin to disgust and even provoke national scandal.[4] This unwillingness to consume horse meat presents a logistical challenge to rewilding projects involving horses. Since publics do not consider horses livestock they are far more resistant to their slaughter or culling than the killing of animals such as cattle (which *are* considered livestock). And, since along with the large herbivores, Britain's large carnivores are also extinct, horses lack any predators which would play a part in regulating their populations. As a result, if humans enrol horses in rewilding projects they are faced with serious challenges regarding the 'management' of horse numbers, risking public controversy if horses are slaughtered for human consumption or culled even if not for human consumption.

Although aurochs are extinct, they are the progenitor of domestic cattle. Therefore, while aurochs are not available for reintroduction/rewilding, their genes survive in domestic cattle (Achilli, et al., 2008). Therefore, domestic cattle can, arguably be viewed, and utilised, as analogues for the aurochs. In some ways, cattle are straightforward in this respect – since there are no surviving wild cattle, all cattle are domestic, meaning that, theoretically at least, they have a less complicated association with categories of wildness and domestication than deer and horses (see DeSilvey and Bartolini, 2018). This is, however, complicated by *back breeding* programmes which seek to reveal the latent similarity between aurochs and domestic cattle. Back breeding (or breeding back) is a method of 'de-domestication' which attempts to recreate an extinct subspecies by artificially selecting for its genes which are still present in the gene pool (Gamborg et al., 2010). This was attempted in relation to cattle/aurochs by the programmes which created Heck cattle in the 1920s (see Lorimer and Driessen, 2016 for a full explanation) and is again being attempted in the current Tauros Programme (Tauros Programme, 2012). While the aim of such projects is to de-domesticate a domestic species in order to recreate a wild species, the human intervention, in the form of selective breeding, means that the cattle produced are, essentially, another domestic breed (there are interesting points of comparison here with wild and domestic cats which are discussed in Chapter 7).

Categories of wild and domestic are further complicated by the position which domestic cattle, including Heck cattle, occupy once there are deployed in rewilding programmes. In Britain, the cattle used tend to be hardy, native breeds, perhaps most notably Highlands and Black Galloways. Heck cattle have been used previously and, while at time of writing this is no longer the case, their involvement illustrates some of the challenges involved in rewilding with domestic cattle, particularly a breed deliberately developed to have 'wild' traits such as aggression. Precisely because cattle are a domestic species, they are subject to policies to protect human and animal health, perhaps most significantly in this context testing for bovine tuberculosis (bTB or TB), although regulations requiring the removal and disposal of fallen stock are also significant. Therefore, even when domestic cattle are living relatively independently from humans, and acting as analogues for aurochs, they must still be rounded up for bTB testing at intervals ranging from six months to four years, depending on the level of risk in the area (Animal and Plant Health Agency, 2020). Bovine tuberculosis testing can be challenging even when dealing with highly managed herds. When dealing with herds which are subject to very little human management, bTB testing can become dangerous since cattle are intractable and unused to being handled. This situation is exacerbated when breeds such as Heck cattle are chosen by rewilding projects for their 'wild' traits which, while attractive from the rewilding point of view, are problematic from a management perspective. This tension between wild and domestic traits and wild and domestic status was illustrated very clearly during an interview with a rewilding practitioner[5]:

> We used to have Heck cattle here, we've got one Heck cow now which is the only surviving Heck in the UK but basically its herd mates all became sausages at the end of last year. Initially they were okay, some were more psychotic than others so they were killed but there were some nice ones left behind and we found that actually, if you treated the nice ones like domestic cows and just kept them like domestic cows they're absolutely fine. And then a couple of years ago we basically put the ring fence around the wilding area and let the Heck in there just to be basically, but being domestic cattle they still had to be rounded up for TB testing and all the rest of it, which was a bit of a palaver, but it was just about manageable. Once they were in the wilding area and were left to forage and roam through the woodlands they went completely wild, they all got very dangerous and then when it came to TB testing there were some very near misses and it's just like look, you know, if we keep these someone's going to get killed eventually during TB testing. And because you cannot keep cattle in the UK if you can't TB test them, they all got to be shot basically. So that's basically been a very hard lesson in the practicalities of using some of these sorts of wild analogues for rewilding.

This quote clearly demonstrates the frictions that arise when cattle are simultaneously subject to competing biopolitical modes of agriculture and rewilding. It also demonstrates, however, that, in other ways, cattle can be relatively straightforward to govern as a *result* of being governed by these two modes. By 'all becoming sausages' the Heck cattle who were involved in this rewilding project then became involved in the agrifood system. This highlights the potential for overlap between rewilding and extensive farming and also the way that cattle, since viewed a as production animal, can be lethally managed by rewilding projects by reverting to the biopolitics of agriculture. Slaughter of cattle as part of human food production, even if it is to manage the number of cattle involved in a rewilding project, is unlikely to be met with public disapprobation: indeed, not only can 'surplus' cattle from rewilding projects enter the food chain, they can often be marketed as a premium product (e.g., 'wild' meat (Knepp Wildland, 2022). Thus, the deployment of cattle in rewilding projects avoids the (potentially) more problematic negotiation which would be required with species such as deer and horses which are not necessarily considered a food animal and which occupy a more liminal space with regard to wildness and domesticity.

The liminal cattle of rewilding

As outlined in the previous section, domestic cattle in Britain are subject to legally mandated management interventions, particularly in relation to disease surveillance. If wild cattle were present in Britain they would not be subject to these same regulations and interventions. This distinction is largely based on One Health considerations: domestic cattle are production animals whose products (meat and milk) will enter the human food chain. The health of domestic bovids therefore has far greater potential to impact human health than does the health of wild bovids. To a certain extent, however, the distinction is also a result of our different attitudes towards and, therefore, treatment of wild and domestic animals. We have a duty of care to domestic animals which we do not owe to wild animals (Donaldson and Kymlicka, 2011). Thus, not only do we have a responsibility to intervene in domestic bovine lives in order to protect human health and wellbeing, we also have a duty to protect the health and welfare of the bovids involved.

Having argued in the previous section that, on one level, cattle occupy as less complicated status with regard to wildness/domesticness than deer or horses, cattle's deployment in rewilding projects blurs their wild/domestic status, leaving them in a liminal state in which they are both wild and domestic and neither wild nor domestic. This liminality does not alter their status to the extent that they are exempt from bTB testing regulations, nor does it exempt their 'keepers'[6] from their obligations under the Animal Welfare Act 2006 and the associated regulations for the welfare of farmed animals.[7] The effect of such regulations is that even when cattle are involved in rewilding projects and living in a way which, up to a point, resembles the way a wild bovid might live, it is not legally acceptable to withdraw human management entirely or to relinquish responsibility for the cattle's health and welfare. This stipulation has meant that a situation such as transpired at the Oostvaardersplassen in the Netherlands has never occurred in Britain.

The Oostvaardersplassen is a rewilding project not far from Amsterdam. The land of what is now the Oostvaardersplassen was reclaimed from the sea for development but, due to economic constraints, the planned development never took place and the land was left for nature. Geese colonised the area and it became recognised as an important habitat for waterbirds. In order to maintain this habitat, and to prevent vegetation succession resulting in the scrubbing up of the reedbeds which the geese utilised, cattle, deer and ponies were introduced to graze and browse (Keulartz, 2009; Lorimer and Driessen, 2013; Lorimer and Driessen, 2014). These animals were considered 'wild' by those involved in the rewilding project despite the fact that they could not leave the rewilding area, since the Oostvaardersplassen is bound either by water or by fences (Keulartz, 2009; Lorimer and Driessen, 2013; Lorimer and Driessen, 2014). As a result of being considered wild they did not receive supplementary feeding or veterinary intervention and, when they died, their bodies were left *in situ* to provide food for scavengers and necrophages and form part of the decomposition cycle.

Following their introduction, the populations of cattle, deer and ponies increased. Population growth was sustained until a particularly hard winter in 2004/2005 resulted in a shortage of food for the, by now, large number of animals. The custodians of the project deemed that this was the result of the number of animals exceeding the carrying capacity of the land and that the resulting starvation was a 'natural process'. Therefore, in line with their rewilding ethos of non-intervention, they did not provide supplementary food for the animals (despite the fact that the animals were not able to leave the area to seek food elsewhere). As a result, a significant proportion of the animals died and, since their bodies were left *in situ*, the Dutch, and later international, media provided publics with scenes of starving and dead animals: public and media outcries ensued (Keulartz, 2009; Lorimer and Driessen, 2013; Lorimer and Driessen, 2014). Some people took direct action to feed the animals, either cutting fences to access the Oostvaardersplassen or throwing food over them. Others took legal action which, ultimately, resulted in those involved in the management of the Oostvaardersplassen changing their policy on intervention so as to maintain their social licence to operate (Keulartz, 2009; Lorimer and Driessen, 2013; Lorimer and Driessen, 2014).

What the case of the Oostvaardersplassen demonstrates is that perceptions of when animals are 'wild' and when they are 'domestic' can vary, as can attitudes as to when, or if, humans have a responsibility to intervene in their lives. While, to date, this has not been so starkly illustrated in Britain, this case highlights the potential for controversy surrounding the care of animals involved in rewilding projects, particularly when there is room for interpretation as to their wild/domestic status. At this stage, cattle involved in rewilding projects in

Britain are treated differently from their more intensively farmed counterparts, from wild animals, and from cattle involved in other rewilding projects. This different treatment manifests as the amount of agency the cattle are granted and as the amount of intervention they receive, specifically in relation to feeding and (veterinary) care. As a result, different ways of living with cattle arise and different cattle–human relationships emerge with the cattle involved in rewilding projects. This requires new modes of biopolitics since the governance of these animals does not align with established biopolitical modes for the governance of cattle. This aligns with suggestions from scholars that rewilding requires, and creates, new modes of biopolitics for its interactions with the other-than-human animals involved (Lorimer and Driessen, 2013, 2016).

Biopolitics and the biopolitics of conservation and rewilding

Foucault's (1978) notion of biopolitics is concerned with the regulation of, and exercise of power over, living things, including the administration, optimisation and multiplication of life, seeking to exert a positive influence that will ensure, sustain, and multiply it. The exertion of power in an attempt to order life, albeit with benign intent, subjects life to 'precise controls and comprehensive regulations' (Foucault, 1978). In its attempts to regulate and sustain life, biopolitics deals with populations, thus, when applied to other-than-human animals, biopolitics focuses on species rather than individuals, intervening in processes such as birth, death and illness at a species level to 'make live and let die' (Foucault, 1976). Making live and letting die emerges as power evolves from sovereignty to biopolitics (Foucault, 1976). Whereas a sovereign has absolute power over their subjects, biopolitics makes more allowances for the agency of those governed. Sovereignty holds the power to *'take life or let live'*, a notion which centres on *death* and the ability to evoke or postpone it (Foucault, 1976). Biopolitics meanwhile is a strategy for governing *life* (Rabinow and Rose, 2006), so *'makes live and lets die'*, compelling or denying life (Foucault, 1976).

Conventional conservation, with its target-driven approach and its concentration on specific species and habitats, focuses very strongly on the make live, let die ethic (van Dooren, 2014). For example, species which are considered desirable, and are targeted for conservation, can be *made live* by ensuring that the individuals which constitute the species exist or continue to exist by, for example, establishing breeding programmes or providing supplementary feeding (van Dooren, 2014). On the other hand, species which are *not* considered desirable,[8] and are not therefore targeted for conservation, are excluded from such care programmes and 'let' die, through sacrifice or neglect (van Dooren, 2014). The sacrifice or neglect may be intentional or unintentional and even conscious or unconscious; individuals may be compelled to perform roles which expose them to risk of death, or entire species may be allowed to perish due to lack of intervention if conservation efforts are directed elsewhere (van Dooren, 2014).

Rewilding's 'hands-off approach' (Corlett, 2016) aligns very strongly with this 'let die' ethic. This has earned criticism regarding its lack of care for the other-than-human animals involved as occurred, for example, at the Oostvaardersplassen. In addition, rewilding's uniform application of the let die approach earns further criticism from those who fear that vulnerable species will fare less well as a consequence of rewilding than in conventional conservation regimes (Sandom et al., 2018). Meanwhile, rewilding's open-ended approach and avoidance of targets or goals, means that it has no commitment to making live. Because rewilding does not align with the make live element of established biopolitics, it maps somewhat unsatisfactorily onto current modes of biopolitics as a whole. Rewilding, therefore, necessitates the development of new modes of biopolitics to recognise the human–animal relations involved.

Scholars have started this work in relation to Heck cattle in the Oostvaardersplassen: they identify four prevalent, conventional biopolitical modes relating to cattle (agriculture, conservation, welfare and biosecurity) and offer a biopolitics of rewilding as an addition (Lorimer and Driessen, 2013). Since this biopolitical mode proposed is specific to the governance of Heck cattle involved in the Oostvaardersplassen rewilding project, scholars note that there is further scope to explore the multiple modes of living with and relating to the other-than-human animals involved in rewilding and the 'connections, frictions and compromises between modes of relating' (Lorimer and Driessen, 2013, p. 257). The biopolitical modes proposed here explore those multiple modes of living and relating. Furthermore, since the research on which the biopolitical modes are based examined different breeds of cattle (and other species) in different rewilding projects it is possible to extend earlier analyses (e.g., Lorimer and Driessen, 2013) and identify the *multiple* biopolitical modes within rewilding, accepting that there is not a single biopolitical mode of rewilding (c.f. Biermann and Anderson, 2017) who note that there is not a single biopolitics of conservation). It should be noted, however, that the multiple biopolitical modes are not clear and distinct (though for the sake of clarity they are discussed here as discrete logics), rather, there are 'overlapping and contradictory logics and techniques' (Biermann and Anderson, 2017). Furthermore, cattle (and other animals) do not necessarily fit within one biopolitical mode, they move between them and can even occupy one or more mode simultaneously, complicating biopolitical governance.

In relation to the emerging biopolitics of rewilding, it is important to clarify its position with respect to life and the conventional biopolitical logic of make live. Rewilding's logic of increasing other-than-human agency and the avoidance of coercive interventions means that a more accurate interpretation of its logic in relation to life would be *let* live rather than *make* live. This is revealed in the biopolitical mode of 'self-determining agents', as will be discussed. While rewilding does not align with make live ideologically, the domesticated nature of rewilding in Britain means that biopolitical modes there do tend to slip into the conventional logic of make live (especially where domestic other-than-human animals are involved), despite such coercive logic being, at least in principle, antithetical to rewilding. Thus, the other biopolitical modes identified here (expendable objects, animal machines/human proxies and analogue species) reveal a reliance on the conventional make live logic. The significant difference between the logics of self-determining agents and the logics of the other biopolitical modes of rewilding is due to the degree of agency afforded to the other-than-human animals in question.

Agency[9] is 'the capacity to contribute to the future; the ability through action, interaction or deliberate inaction to change the outcome of events' (Rees, 2017, p. 9). A very important aspect of agency with respect to rewilding is its relationship with wildness, with wildness being 'the autonomy of the more-than-human world where events, such as animals moving about, plants growing, and rocks falling occur largely because of their own internal self-expression that is independent of civilized forces' (Woods, 2005, p. 177). With respect to the biopolitics of rewilding then, agency can be seen as the self-expression of other-than-human actors, *independent* of civilized forces, and it is wildness which gives actors this agency, or the ability to express it.

Another important aspect of agency with respect to rewilding is that it can be exercised at individual, species and interspecies (i.e., ecosystem assemblage) levels. Given rewilding's focus on ecosystem function, the balance of other-than-human agency in rewilding projects theoretically rests at interspecies level rather than at the level of the species or the individual, as is more common in the biopolitics of conventional conservation, domestication or captivity. In practice, however, this may not always be the case. For example, again in the case of domesticated rewilding in Britain, its biopolitical logics can slip into the established logics of domestication, captivity or conventional conservation, which centre on the agency of individuals or species, rather than the logic more aligned with the tenets of rewilding, which centres on interspecies agency. The biopolitical modes of rewilding in Britain, including the location of agency, are outlined in the following sections and summarised in Table 5.1.

Species as expendable objects

The biopolitical mode of rewilding which is most similar to conventional biopolitical modes (particularly that of agriculture) and aligns with established make live let die logics, is that of species as expendable objects. While cattle are seen as important actors in rewilding in Britain for the roles they perform and the niches they fill (as will be discussed in subsequent sections) they are also, at times, considered expendable, either in terms of their presence not being required on a permanent basis or, more seriously, in terms of their lives being treated as dispensable. The cattle involved in rewilding at the Avalon Marshes and Wild Ennerdale were exposed to considerable risks as part of their enrolment in the rewilding projects, largely because of the

Table 5.1 Typology of the biopolitical modes of the companion species of rewilding

Description	*Expendable objects*	*Animal machines/human proxies*	*Analogue species*	*Self-determining agents*
Aim	Species are human subjects.	Species perform tasks on behalf of humans, often acting as tools or machines.	Species act as analogues for extinct/extirpated species, filling an ecological niche.	Species are at liberty to express their agency.
Location of agency	Species or individual level.	Species or individual level.	Interspecies or species level.	Interspecies or species level.
Logics of life	Make live: life is not valued, these species are considered disposable.	Make live: life is valued from an instrumental perspective for the role that these species play. Individuals and even species are not necessarily valuable if other individuals or species exist which could perform the same role.	Make live: life is valued for what it replaces rather than for its own sake. The species as a whole is valued more highly than the individual.	Let live: life is valued as part of a functioning ecosystem.
Logics of death	Let die: death (or presence) is irrelevant. These species are seen as dispensable.	Let die: death of the individual and species are to be avoided if possible, but it is not necessarily critical to do so.	Let die: death of the species is to be avoided, as is the death of individuals if possible, although individual deaths are not necessarily considered important, especially if the species is not threatened.	Let die: death is undesirable at a species level but acceptable at an individual level. When death occurs, it is 'natural' and bodies are left in situ.
Example	Low-intensity grazing by Highland Cattle in the Avalon Marshes.	Red Devons performing conservation grazing in the Avalon Marshes.	Black Galloways performing naturalistic grazing in Wild Ennerdale.	Heck cattle in the Oostvaardersplassen.

Description	*Expendable objects*	*Animal machines/human proxies*	*Analogue species*	*Self-determining agents*
Illustrative quote	'They're seasonal and they'll get taken out. We have contracts with people who bring them in' (i.e., they are dispensable).	'We have livestock grazing here, which is a management tool' (i.e., they are performing a task).	'As soon as we put them [the Black Galloways] in, everything changed in terms of the nature of the sward, we got much more regeneration, we got much richer swards coming through, we've got dung beetles in their cowpats' (i.e., they are filling an ecological niche).	'We basically put the ring fence around the wilding area and let the Heck in there, just to be basically' (i.e., they are at liberty to express their agency).
Level of other than human agency	Ability to express agency is constrained.			Unconstrained ability to express agency.

Source: (after Lorimer and Driessen, 2013).

experimental nature of rewilding and the lack of knowledge regarding how the cattle would fare. While rewilding stakeholders spoke of close surveillance of the cattle to monitor their welfare, little concern was demonstrated in relation to exposing the cattle to risk, viewing it as an acceptable part of the rewilding experiment, thus revealing a view of the cattle as disposable.

Given the experimental nature of rewilding, the risks involved in enrolling cattle in the rewilding of the Avalon Marshes and Wild Ennerdale were somewhat unknown, something which is recognised in the discourse on rewilding (e.g., Nogués-Bravo et al., 2016). This was stated clearly in relation to Wild Ennerdale, where cattle had not been present in the valley for decades. Peggy, a landowner/manager in Wild Ennerdale, explained that when the introduction of cattle was proposed 'everyone said we were crazy because we didn't know what disease there may be or animal health problems on land that's never been farmed for a long time, it was said that we would have trouble with redwater,[10] with ticks'.[11] Other research participants made similar comments about the unknown risks of reintroducing cattle to Wild Ennerdale, and the negative consequences that were predicted. For example, Ross, a conservationist, said 'all the other farmers, said they'd [the cattle] be dead, belly-up in the lake within weeks' which was echoed by Max, a conservation advisor,

> all of the other farming community around him [the farmer responsible for the cattle in Wild Ennerdale] said "that won't work, you can't do it, the animals will be dead, they'll be floating in the lake". While research participants went on to explain that, for the most part, these dire predictions had not been borne out, only Peggy spoke of the care that had been taken to ensure that this was the case, saying that they intended to 'keep a good eye on them [the cattle]', and 'if things were going wrong, we could hopefully spot it and right it

This suggests that, for the other participants at least, these cattle are considered disposable and are wittingly exposed to risks and unfavourable conditions which they acknowledged would be too harsh for less hardy breeds. For example, Niall, a conservationist at the Avalon Marshes, recognised that 'some of the more commercial animals, if you tried them on this land they wouldn't do so well'. These conditions explain the use of hardy, native breeds in rewilding projects, such as the Highland cattle in the Avalon Marshes.

As well as being exposed to harsh conditions, the disposable nature of these Highland cattle was evident in the way that they were only temporary members of the Avalon Marshes ecosystem. Octavia, a conservationist, explained that the cattle are 'not a complete part of the process within our site ... they come and go'. This point was reinforced and extended by Maeve, another conservationist at the Avalon Marshes, who explained that the grazers 'are seasonal and they'll get taken out, we have contracts with people who bring them in, they just get used on different sites and [they get taken out] when the gorse isn't growing and they've done their bit'. This comment highlights the expendability of the grazers, in that once that have 'done their bit' their presence is no longer required or even desired.

Animal machines/human proxies

Within rewilding projects, other-than-human animals are often deployed as 'proxies' for humans, performing ecological restoration on their behalf (von Essen and Allen, 2016). Other-than-human animals in this biopolitical mode are 'animal machines/human proxies', being viewed by humans teleologically, as a rewilding 'tool'. Other-than-human animals subject to this logic are deployed to perform a certain function and are therefore valued for their instrumental purpose rather than their intrinsic qualities. Again, the established biopolitical logic of make live, let die is at play here; since other-than-human animals are viewed as machines, if another individual, or even species, could perform the same role then the life of an individual animal, or even an entire species is not necessarily valuable. Other-than-human animals in this mode have more agency than those which are considered 'expendable objects' but they are allowed to express this agency only by human consent. Indeed, this agency is channelled by humans in order to achieve the desired outcome.

The liminality of other-than-human animals in this biopolitical mode is stark since, while they are exerting their agency, they are not exerting it on their own behalf but on behalf of humans. Crucially, because their agency is being exploited by humans, if it ceases to perform the desired function it ceases to be useful and humans are liable to perform *'ad hoc'* interventions to curtail or redirect that agency (von Essen and Allen, 2016). Scholars have described the uneasy balance which animals in this mode must strike between being 'wild but not too wild' as 'Goldilocks conditions' (Von Essen and Allen, 2016). The term Goldilocks conditions is usually applied to a situation where 'ideal' conditions exist. In the case of animal machines/human proxies, Goldilocks conditions with respect to expression of other-than-human agency are viewed from the human perspective and maintained through management interventions. For the other-than-human animals involved, however, the conditions are far from Goldilocks. Perhaps 'bear conditions' would be a more apposite term for the difficult position they occupy in which they fluctuate between being too wild (from the human perspective) until a human 'correction' is imposed, and not wild enough (from the other-than-human animal perspective) once the correction has taken place and they are faced with constraints on their agency.

The case of cattle this mode is clearly illustrated via conservation grazing, again with cattle as the tool of choice for rewilding. Conservation grazing entails the use of herbivores to maintain habitats in a particular condition, suppressing vegetation succession through grazing pressure (Hodder et al., 2005). Conservation grazing is, as the name suggests, primarily a conservation technique rather than a rewilding technique although it can be used as part of rewilding strategies. The use of Red Devon cattle as conservation grazing tools was highlighted by Alan, a conservationist, at the Avalon Marshes when he said, 'we have livestock grazing here, which is a management tool'. Furthermore, the way cattle have replaced machines, and become machines themselves, was illustrated by a landowner and manager at the Avalon Marshes who acknowledged that 'we do grazing with livestock whereas it would be done with machinery maybe'.[12] The extent to which, in performing their conservation grazing role as tools or machines, cattle are human proxies was illustrated by Octavia, a conservationist, who said that cattle 'do a good job for us' exemplifying the way that cattle are being recruited to act on behalf of humans.

Species as analogues

The role of analogues for other species requires another mode of biopolitics to accommodate the greater level of other-than-human agency involved. As discussed above, cattle can be deployed as analogues for the aurochs (and, less directly, other absent herbivores), acting as a dynamic force and disturbance factor within the landscape through naturalistic grazing and poaching the ground. Naturalistic grazing differs from conservation grazing in that, rather than being a means to an end in creating or maintaining a desired habitat, naturalistic grazing is an end in itself: the emphasis is on reinstating grazing as a natural process as part of restoring ecological function (Hodder et al., 2005). Despite cattle's suitability to perform this role, it should be noted that because they are a domestic species and because, as will be explained, they ultimately play a role in the agrifood system as well as performing their role in rewilding, the function they perform is, arguably, not a 'natural' process but an artificial or cultural one, hence referring to *naturalistic* rather than *natural* grazing.

Rewilding stakeholders at the Avalon Marshes and Wild Ennerdale identified cattle as analogues for other species either more or less explicitly: Max, a conservation advisor at Wild Ennerdale, compared the role of the Black Galloways there directly to that of the aurochs. Peggy, a land owner/manager in the Ennerdale Valley, referred to the cattle's ecological function as large herbivores, and Vanessa, another Ennerdale land owner/manager, spoke of the cattle as filling an ecological or trophic niche within the ecosystem. It is cattle's size and grazing habits which make them seen as appropriate analogues for the aurochs. Size is important in creating 'regeneration niches', as Max explained: 'they're [the cattle] a bit heavier [than sheep] so their footprints become regeneration niches'. Max also explained that cattle's grazing habits were important since 'they don't selectively graze, they just rip out everything with their tongues and their mouths and they take coarse stuff as well as fine stuff'. This was echoed by Ross, a conservationist in Wild Ennerdale, who explained that the way cattle graze, 'wrapping their tongues round [forage] and ripping, so you get these areas where there can be regeneration', replicates the grazing and browsing habits of the aurochs, thereby filling that ecological niche.

In acting as analogues for aurochs, cattle in this mode have a considerable degree of agency. This agency is, however, surveilled and curtailed by humans. This was illustrated by the fact that, while participants spoke of cattle as 'doing their own thing' such statements were usually prefaced with the notion that humans 'let' or 'allowed' them to do so, thereby implying that cattle were able to exercise their agency only by human consent and within certain limits. The ultimate curtailment of bovine agency is enacted through their slaughter. This is because in this mode (and indeed the modes of expendable objects and animal machines/human proxies) while the cattle are involved in rewilding they are also, simultaneously, part of extensive farming systems in which they remain production animals. The capacity of cattle to inhabit agricultural and rewilding 'spaces' simultaneously is, as outlined earlier, one of the factors which make them attractive to those involved in rewilding projects. The ability to slaughter cattle as part of the established agrifood systems means that 'superfluous' cattle can be managed in a way which is unlikely to cause public controversy, as opposed to the culling of wild, charismatic or non-food animals, or allowing animals to starve to death, both of which provoke more controversy than animal slaughter as part of food production. It is also important to note the distinction between the slaughter of cattle as a form of 'management' and the biopolitical mode of species as expendable objects. While slaughter is used as a means of controlling cattle in the modes of animal machines/human proxies and analogue species, it is not done because the animals are seen as expendable objects but because, in the absence of apex predators, as in Britain, there are no carnivores to exert predatory pressure on cattle. Human intervention therefore replaces this missing ecological process.

Species as self-determining agents

The final biopolitical mode identified, in which other-than-humans exercise the highest level of agency, is that of self-determining agents. This mode is still emerging in rewilding in Britain, certainly with respect to cattle since, as a domestic species, their agency is, to at least some extent, subject to human intervention. Nonetheless, elements of its logic were evident in relation to the cattle involved in the rewilding of Wild Ennerdale. Peggy described how, even as a domestic species, the cattle are 'given a lot different parameters, and a lot different area to do what [they] want, rather than to do what their keeper decides'. Peggy went on to describe how the cattle exerted their agency within these new, somewhat 'de-domesticated' (Lorimer and Driessen, 2013) parameters, emphasising their freedom to roam the Ennerdale Valley: 'they could roam, they could do their own thing, whether it be under trees, whether it be in the water, in the river, and in the lake, up high on the fells' and 'they go into the forest in times of heavy rainfall, for shelter, and they'll go higher up on the slopes in times of hot weather, to try and get away from the flies, and get a bit more air, wind'. Similarly, Ross emphasised the cattle's liberty to go where they liked and that, as a result, they were able to express their preferences, saying, 'there's routes that the cattle go, totally favour, and areas they just ignore entirely'. The language used by participants here was different from other comments where the cattle were described as being 'let' or 'allowed' to behave in a certain way. In these examples, the cattle are free to express their agency without needing permission from humans, illustrating their status as somewhat self-determining agents.

While the cattle in Wild Ennerdale are allowed to be self-determining in some respects, they are still subject to human intervention in other respects, as indeed are all other cattle in Britain. It is, therefore, necessary to look outside Britain for an example of cattle in this biopolitical mode and return to the Heck cattle in the Oostvaardersplassen. While the case for these cattle being entirely self-determining agents is imperfect,[13] they nonetheless provide an illustrative example of how this biopolitical mode might manifest itself in relation to cattle involved in rewilding projects. Perhaps self-evidently this mode can be applied more readily to the wild companion species of rewilding although this is still not necessarily straightforward. For example, where species are seen to conflict with human interests or values, humans are likely to intervene to constrain the agency even of wild species. This is currently evident in Britain in the way that lethal control of beaver is permissible in Scotland. Furthermore, were large carnivores to be reintroduced to Britain, it is highly unlikely that such species would be permitted to be self-determining agents: it is likely that humans would insist on the right to curtail their agency.[14]

The fundamental difference between the biopolitical logic of self-determining agents and the other modes is that self-determining agents are let live rather than made live. The logic of let live is not coercive in the same way as make live since animals are not *made* or *forced* to live but *let* live. Indeed, in some cases, they are *allowed*, or even *enabled*, to live, perhaps more in line with a sense of 'flourishing' (Haraway, 2003; Tsing, 2012). With the exception of the mode of species as self-determining agents, however, the biopolitical modes identified here demonstrate little evidence of the 'radical tolerance' of other species (Campbell, 2006) or the 'shift from a biopolitics as a *control over life* to a biopolitics of *living with*' (Lorimer and Driessen, 2013, emphasis added) which rewilding proposes. Whereas the biopolitical logic with regard to self-determining agents can be described as 'let live and let die' the other biopolitical modes (analogue species, animal machines/human proxies, and expendable objects) adhere to the classic Foucauldian logic of 'make live and let die' (Foucault, 1976).

The biopolitics of rewilding

The four biopolitical modes outlined here (expendable objects, animal machines/human proxies, analogue species and self-determining agents) exemplify the new forms of biopolitics which are emerging from rewilding. Given that rewilding is still, relatively, new as a conservation approach and is still evolving, it is likely that the biopolitics of rewilding will, likewise, continue to evolve. Nonetheless, the identification of these modes and their enactment in rewilding projects illustrates the current state of rewilding biopolitics, at least as they apply in Britain, and it is reasonable to propose that the same, or similar, logics are at work in relation to rewilding in other countries.

While the modes identified here represent the current state of rewilding biopolitics, the intersection of life and politics will continue to change and shift as new biopolitical spaces created by rewilding emerge. It is possible that modes of biopolitics will emerge which more closely resemble ways of living with other-than-human life rather than controlling it (Lorimer and Driessen, 2013). It is, however, equally possible that more restrictive modes of biopolitics emerge if humans perceive the species involved in rewilding as conflicting with their interests or values.

While cattle have been used here to illustrate the biopolitical modes proposed, the modes are equally applicable to other species involved in rewilding. This will be demonstrated in the following chapters in discussing the rewilding of the red kite (*Milvus milvus*) and the European wildcat (*Felis silvestris*) in Britain including, *inter alia*, the way these species are perceived and governed. Two points, however, cannot be overstated, and should be reiterated here. First, these emerging forms of biopolitics entail significant frictions and compromises as new human–animal relationships are negotiated. Secondly, the biopolitical modes themselves are complex and overlapping, with animals moving between them or occupying more than one mode simultaneously. This will continue to become evident in the following chapters.

Notes

1 Analogue species can serve to replace extinct species but also extant species from other ecosystems or regions where, for example, changing environmental conditions or human regulations would make translocation of those species unfeasible.
2 Use of the term 'tool' suggests that these cattle are viewed teleologically – while this is sometimes true it is not always the case, as is discussed later in this chapter. Tool is used here as a catch-all term rather than as a specific descriptor.
3 This was illustrated in an interview with Stephen, an Ennerdale Valley landowner, who saw a 'contradiction' in the fact that roe deer within the Wild Ennerdale boundary were shot, with 'stalkers blowing their heads off all over the place', despite the Wild Ennerdale partnership purporting to be reducing human intervention. Stephen saw this ad hoc intervention as inconsistent, saying 'where do we draw the line between "wild" and "management"? If we've got loads of people in there removing undesirable species and replacing them with desirable species, that's not rewilding. That's not natural processes. That's management. That's farming or forestry. That is doing something to deliver an outcome'.
4 The 2013 'horsemeat scandal' in Britain resulted when products labelled as containing beef were found to also contain horse meat. While part of the scandal was

related to the mis-labelling/mis-selling of food it was heightened by people's reaction to having unwittingly consumed horse (Lawrence, 2013). Another scandal, 'let them eat horse', was provoked when Princess Anne, a senior member of the British royal family and well-known as an equestrian, suggested, in response to welfare concerns regarding free-living horses, that they might receive better care if they were thought of as meat animals. The comments were interpreted as callous in depicting horses as 'disposable livestock' (Weaver, 2013) and hypocritical given that Princess Anne was considered a horse lover.

5 The participant's pseudonym is not used here since the detail in this quote could potentially compromise their anonymity. If their pseudonym were used, it would be also possible to attribute their other comments to them.

6 'Keepers' is used here since it is the terminology used in the legislation and guidance. The extent to which animals involved in rewilding projects are 'kept' is, of course, interesting and while this is not yet significantly tested or challenged in Britain it was certainly tested with regard to Heck cattle in the Oostvaardersplassen.

7 Again, 'farmed' is used here since it is the terminology used in the legislation and guidance. Three things are of note here. First, the lack of rewilding specific legislation, meaning that cattle (and other animals) involved in rewilding projects still fall into the category of farmed animals from the point of view of legal governance. Second, and following on from the first point, in Britain, even when involved in rewilding, cattle are still very much considered a domestic, farmed animal. This is partly a response to legislation and partly a response to attitudes to cattle, both of which contribute to the 'domestication' of rewilding and the choices made in relation to animals in rewilding projects (e.g., using domestic analogues and maintaining human management of them which, in turn, feeds into human perceptions). Third, agri-environmental legislation in the UK is devolved, meaning that it is governed separately by the national governments of England, Scotland, Wales and Northern Ireland. While broadly similar in relation to animal welfare there are technical differences between the legislation in each nation.

8 It should be noted that not being considered desirable is very different in conservation terms from being considered *un*desirable. Species which are specifically considered undesirable are often subject to targeted persecution rather than 'simply' (benign) neglect.

9 Agency and autonomy are often used interchangeably in rewilding discourse and, while a literature based on 'autonomy' is emerging, agency if used preferentially throughout this book since it avoids some of the specific philosophical connotations which autonomy possesses: naturally 'autonomy' is retained if it appears in referenced work or in contributions from research participants.

10 Redwater fever or babesiosis is a tick-borne disease which affects cattle and causes, *inter alia*, haemoglobinuria, hence the name 'redwater' (NADIS, 2020).

11 The participant's pseudonym is not used here since the detail in this quote could potentially compromise their anonymity. If their pseudonym were used, it would then become possible to attribute their other comments to them.

12 The participant's pseudonym is not used here since the detail in this quote potentially compromises their anonymity. If their pseudonym were used it would be possible to attribute their other comments to them.

13 The Heck cattle of the Oostvaardersplassen are not a perfect example since their movement is constrained by fencing and because, following the controversy over their starvation in 2004/5, attempts are now made to avoid animals starving to death during winters by culling them before this can occur, thus further curtailing their agency.

14 While at the time of writing, there are no immediate plans to reintroduce large carnivores such as the Eurasian wolf (*Canis lupus*) or the Eurasian lynx (*Lynx lynx*) to Britain, debates over whether to do so have been going on for many years and look set to continue.

References

Achilli, A. et al. (2008) 'Mitochondrial genomes of extinct aurochs survive in domestic cattle', *Current Biology*, 18(4), pp. 157–158.

Animal and Plant Health Agency (2020) *Bovine TB testing intervals*. Available at: https://www.gov.uk/guidance/bovine-tb-testing-intervals (Accessed: 21 June 2025).

Biermann, C. and Anderson, R. M. (2017) 'Conservation, biopolitics, and the governance of life and death', *Geography Compass*, 11(10), pp. 1–13.

Campbell, T. (2006) '"Bios," immunity, life: The thought of Roberto Esposito', *Diacritics*, 36(2), pp. 2–22.

Corlett, R. T. (2016) 'Restoration, reintroduction, and rewilding in a changing world', *Trends in Ecology and Evolution*, 31(6), pp. 453–462.

DeSilvey, C. and Bartolini, N. (2018) 'Where horses run free? Autonomy, temporality and rewilding in the Côa Valley, Portugal', *Transactions of the Institute of British Geographers*, 44, pp. 94–109.

Donaldson, S. and Kymlicka, W. (2011) *Domesticated Animal Citizens in Zoopolis: A Political Theory of Animal Rights*. Oxford: Oxford University Press.

Foucault, M. (1976) *'Society Must Be Defended' Lectures at the College de France, 1975–76*. New York: Picador.

Foucault, M. (1978) *The Will to Knowledge*. London: Penguin.

Gamborg, C. et al. (2010) 'De-domestication: Ethics at the intersection of landscape restoration and animal welfare', *Environmental Values*, 19(1), pp. 57–78.

Haraway, D. (2003) *The Companion Species Manifesto: Dogs, People and Significant Otherness*. Chicago: Prickly Paradigm Press.

Hodder, K. H. et al. (2005) *Large Herbivores in the Wildwood and in Modern Naturalistic Grazing Systems*. Peterborough. Available at: publications.naturalengland.org.uk/file/93025

Keulartz, J. (2009) 'Boundary work in ecological restoration', *Environmental Philosophy*, 6(1), pp. 35–56.

Knepp Wildland (2022) *Wild Range Meat*. Available at: https://knepp.co.uk/browse/wild-range-meat/ (Accessed: 21 June 2025).

Lawrence, F. (2013) 'Horsemeat scandal: the essential guide', *The Guardian*, 15 February. Available at: https://www.theguardian.com/uk/2013/feb/15/horsemeat-scandal-the-essential-guide

Linnell, J. D. C. et al. (2015) 'Framing the relationship between people and nature in the context of European conservation', *Conservation Biology*, 29(4), pp. 978–985.

Lorimer, J. and Driessen, C. (2013) 'Bovine biopolitics and the promise of monsters in the rewilding of Heck cattle', *Geoforum*, 48, pp. 249–259.

Lorimer, J. and Driessen, C. (2014) 'Wild experiments at the Oostvaardersplassen: Rethinking environmentalism in the Anthropocene', *Transactions of the Institute of British Geographers*, 39(2), pp. 169–181.

Lorimer, J. and Driessen, C. (2016) 'From "nazi cows" to cosmopolitan "ecological engineers": Specifying rewilding through a history of Heck cattle', *Annals of the American Association of Geographers*, 106(3), pp. 631–652.

NADIS (2020) *Babesiosis (Redwater fever)*. Available at: https://www.nadis.org.uk/disease-a-z/cattle/redwater-fever/ (Accessed: 25 June 2025).

Nogués-Bravo, D. et al. (2016) 'Rewilding is the new pandora's box in conservation', *Current Biology*, pp. 87–91.

Rabinow, P. and Rose, N. (2006) 'Biopower today', *BioSocieties*, 1, pp. 195–217.

Rees, A. (2017) 'Animal agents? Historiography, theory and the history of science in the Anthropocene', *BJHS Themes*, 2, pp. 1–10.

Sandom, C. J. et al. (2018) 'Rewilding in the English uplands: Policy and practice', *Journal of Applied Ecology*, 56, pp. 266–273.

Tauros Programme (2012) *The Tauros Programme*. Available at: https://www.taurosproject.com/ (Accessed: 25 June 2025).

Tsing, A. (2012) 'Unruly edges: Mushrooms as companion species', *Environmental Humanities*, 1(1), pp. 141–154.

van Dooren, T. (2014) *Flight Ways - Life and Loss at the Edge of Extinction, Nucleic Acids Research*. New York: Columbia University Press.

von Essen, E. and Allen, M. (2016) 'Wild, but not too-wild animals: Challenging Goldilocks standards in rewilding', *Between the Species*, 19(1), pp. 80–108.

Weaver, M. (2013) 'Princess Anne sparks 'let them eat horse' headlines over remarks to charity', *The Guardian*, 15 November. Available at: https://www.theguardian.com/world/2013/nov/15/princess-anne-let-them-eat-horse-headlines

Wischermann, C. and Howell, P. (2018) 'Liminality: A Governing Category in Animate History' in C. Wischermann, A. Steinbrecher and P. Howell (Eds), *Animal History in the Modern City: Exploring Liminality*. London: Bloomsbury Academic, pp. 1–24.

Woods, M. (2005) 'Ecological restoration and the renewal of wildness and freedom' in T. Heyd (Ed.) *Recognizing the Autonomy of Nature: Theory and Practice*. New York: Columbia University Press, pp. 170–188.

6 Of kites and men

Valuing, categorising and controlling kites in Britain

The introduction or reintroduction of species can be an important element of rewilding and can have many motives. While reintroductions or introductions may be deliberate or accidental, this chapter, and indeed this book, focuses on deliberate reintroductions. Reintroductions involve the release of species into areas where they were once native but from which they are now extirpated (JNCC, 1996). For comparison, introductions are the release of species into areas outside their native range (IUCN, 2008), while reinforcement is the release of species into areas within their native range and where populations still exist, in order to supplement that population (IUCN, 2013).

Motives for introductions, reintroductions and reinforcements are very much related to human values and the value humans place on the species in question. Motives include, but are not necessarily limited to, cultural, ecological, economic, ideological, political, social and strategic (after Marino et al., in press). Depending on the motivation, introductions (though not, for obvious reasons, reintroductions or reinforcements) can involve wild or domestic species as analogues for other species, as in the case of the cattle discussed in Chapter 5. Values are, arguably, even more complex than motives and, to borrow from William Morris, can relate to the way a species is seen or perceived as 'useful' or 'beautiful', having a form or function which is either of benefit to humans or which humans can appreciate for its own merits. Values can, therefore, relate to intrinsic as well as instrumental qualities, including aesthetic attributes, or simply the essence of existence. Such values come very much to the fore when humans categorise species. For example, wild species which humans appreciate for their aesthetic qualities and/or which do not compete or conflict with human interests or values are likely to be categorised as 'wild', 'wildlife' or even 'charismatic wildlife'.

Charisma as applied to wildlife is borrowed from its application to humans (which is discussed in detail in Chapter 8). Charisma is a somewhat nebulous concept but, in relation to animals, it is contributed to through the traits of being 'large' (c.f. Hird's (2009) 'big like us'), 'exotic', 'terrestrial' and 'mammalian' (Albert et al., 2018) and affects where animals are positioned in the sociozoological scale (Arluke and Sanders, 1996). Conversely, wild species

DOI: 10.4324/9781003456490-7

which are not considered aesthetically pleasing and/or which compete or conflict with human interests or values are likely to be classified as pests.[1]

Red kites (*Milvus milvus*) offer an excellent case study of the concepts discussed here. Categorised as 'charismatic wildlife', they were reintroduced to England and Scotland, *inter alia,* to support the relict population in Wales. Previously, however, and indeed in some quarters currently, red kites have been classified as pests. This chapter discusses changing human–kite relationships in Britain and how this has affected the way humans care for and about kites, and how they (attempt) to control them, including through feeding. The discussion of biopolitics is continued from Chapter 5, with kites classified as self-determining agents and expendable objects.

Of kites and men: the value of kites

The red kite is native to Britain but was almost entirely extirpated by the 1900s, primarily due to persecution by humans. British conservationists often begin the narrative of the red kite's fortunes in the Medieval period (e.g., RSPB, 2024). This is perhaps an illustration of our anthropocentrism, and that, in telling the story of the kite, what we are actually telling is the story of kite-human relations. Indeed, the Middle Ages are, arguably, the period when significant human intervention in red kite lives began. Scholars record that the kite was widespread in Medieval Britain and a common sight in towns and cities where despite, or even because of, its familiarity it was valued for scavenging and therefore its instrumental role in keeping the streets clean: 'in London their contribution to city hygiene – especially in terms of tidying up some of the more unpleasant wastes from butchery – ensured that both [kites and ravens] were protected by local bylaws' (Williamson, 2013).

Attitudes to the kite changed dramatically, however, with the introduction of the Preservation of Grain Act (1532). The Act classified kites as vermin and offered financial incentives for their killing, something which was done assiduously by those who now ascribed an economic value to kites. This persecution of kites as vermin continued over the centuries, and the development of game bird shooting meant that kites continued to be seen as a pest, perceived as posing a threat to the game birds. As a result of this prolonged and ongoing persecution, the red kite population in Britain was significantly diminished. Rather than inspiring protection, however, the kite's low numbers prompted further persecution, albeit for different reasons. Kites (and their eggs) acquired scarcity value and were highly prized by collectors and taxidermists, valued more as specimens than as living animals: '[egg] collecting was fashionable at the height of the Victorian era, a time when scholars and gentlemen shot and stuffed anything that moved. Egg collecting has been outlawed since 1954, when it became clear that the Victorian hobby was threatening rare birds with extinction' (Overland, 1993). As this quote illustrates, specimen collection (coupled with ongoing persecution to serve game shooting interests) contributed to the kite's continued decline.

It was not until 1903 that moral values predominated, and the kite's intrinsic and existence value superseded the other anthropocentric values and interests related to it (BBPC, 2023). This caused another significant shift in human-kite relations, with humans now categorising kites as charismatic, native wildlife and, theoretically, shifting to a logic of care rather than a logic of violence. Logics of violence still exist in human-kite relations, however, with enduring and emerging logics. Enduring logics are visible in the way that kites are still persecuted as pests, despite the fact that those engaged in the persecution are now committing wildlife crime (Wildlife and Countryside Act, 1981). Emerging logics are evident in the new forms of violence entangled with care, resulting in logics of 'violent care' (*sensu* Van Dooren, 2014). It is these logics of care and violence that are the focus of this chapter, together with human kite relations as they currently stand and as they have been over the 30-year period from the kite's reintroduction to England and Scotland.

While kites were afforded protection in 1903, their extremely low numbers (there was only one known breeding female) and their slow dispersal rate meant that the ability of the population to recover was uncertain, likely to be slow at best and at worst unsuccessful, resulting in their extinction from Wales and therefore Britain as a whole (Carter and Grice, 2006). A decision was made to intervene, translocating chicks from Germany, Spain and Sweden to reintroduce the red kite to England and Scotland. This process involved the capture of fledgling kites from donor sites and their transport to recipient sites, where they were housed until the natural age of dispersal, at which point they were released. Feeding the kites was, obviously, an essential part of the translocation and reintroduction process. While the kites were in captivity, they were entirely dependent on humans to provide them with food to meet their nutritional, physical and behavioural needs. Once the kites were released, humans continued to assume some responsibility for meeting these needs and made food available to kites at the release sites. This feeding had several aims (as will be discussed below), not least of which was to support the kites' survival post-release.

From the point of view of establishing a self-sustaining kite population, the supplementary feeding, and indeed the reintroduction as a whole can be considered a success. Since the reintroduction, kite numbers have increased considerably, and kites have dispersed across Britain. There are now estimated to be 4400 breeding pairs in Britain, representing over 10% of the global population (RSPB, 2024). This is particularly significant since, although red kites are now flourishing in Britain, they are declining in other parts of their range (Birdlife International, 2020). From another point of view, however, the reintroduction programme has not been successful, with the increasing number of kites precipitating another shift in human kite relations and people in some areas (re)classifying kites as pests. This represents the most recent (re)classification in the series of classifications and reclassifications over the course of shared human-kite history, something which applies to kites and other animals as human attitudes and values evolve.

Animal categories and categorising kites

The way humans value other animals affects the way they categorise them, and, in turn, the way animals are categorised affects the way humans value them. For example, in the case of kites, if kites are not valued (and/or if they are seen to conflict with human interests or values), they are likely to be categorised as pests. Since the categorisation of pest carries negative connotations, a feedback loop is created that exacerbates the low value attributed to kites. Conversely, if kites are seen as valuable for their intrinsic and aesthetic attributes, then they are likely to be categorised as charismatic wildlife. This categorisation carries positive connotations, increasing the likelihood that people will continue to value kites.

Another category, that of native (as opposed to non-native, alien or exotic), can also contribute to the categorisation of a species as wildlife versus pest. Because red kites are considered native to Britain, having been recorded there from the Ipswichian Interglacial (Yalden and Albarella, 2018), it is easier for humans to classify them as wildlife than it would be if they were non-native. Being native does not, however, guarantee that an animal will be categorised as wildlife, an animal may still be categorised as a pest even if it is native, as indeed has been the case for kites. That kites have been, and are, categorised variously as 'wildlife' and 'pests' in Britain illustrates the mutability of such categories: these categories are entirely human constructs and may vary according to time, place and socio-cultural factors. This can be seen clearly in relation to kites in the way their categorisation has changed over the course of kite-human history and depending on the humans involved – gamekeepers might classify them as pests while conservationists might classify them as wildlife.

Values are also human constructs and, as a result, are highly anthropocentric, although it could be argued that some (intrinsic or existence value being prime examples) are axiomatic, that is, 'value that exists independently of human valuations' (O'Neill, 1992). Several types of value are ascribed to kites (see Table 6.1 for an overview). We acknowledge that they have an *intrinsic value* in and of themselves, yet, at times, we also ascribe *instrumental or use value* to them, for the contribution they make to keeping streets and roads clean. This instrumental value could also be interpreted as *functional or ecological value* in that the role of scavenging is an important ecosystem process. Very closely related to this, and also related to intrinsic value, is the notion of *biological, biodiversity or conservation value* whereby kites are considered valuable from an environmental perspective for contributing to biodiversity and increasing ecosystem complexity. From a conservation perspective, this value is intensified by *scarcity value*, thus, the kite's rarity in Britain increased its value from a conservation perspective. Indeed, it could even have been said to intensify its *existence value*, the value that we ascribe to something's (continued) existence: if a species is facing extinction, its existence value increases. The kite's rarity also, however, increased its value from the point of view of egg collectors and taxidermists, including its *exchange value*. More

Table 6.1 Types of value ascribed to animals with examples in relation to red kites

Value	*Description*	*Example in relation to kites*
Acquisition	An entity is valued insofar as it can be traded/exchanged for something else, however, the entity may also be considered valuable simply by means of possessing it, with no intention of exchanging/trading it.	Kites, and their eggs, were seen as valuable by taxidermists and egg collectors, partly for the sake of personal collection but also for the price they could command from other collectors.
Aesthetic	An entity is valued for its physical appearance. It should be noted that, perhaps more than others, this value is subjective rather than objective.	Kites are considered beautiful, graceful birds, meaning that humans ascribe to them an aesthetic value, this also contributes to kites being attributed with charisma, which reinforces their value.
Biological/ biodiversity/ conservation	An entity is valued for its contribution to the complexity of an ecosystem. (Note that while this might appear similar to intrinsic value in that something is essentially considered inherently valuable due to its biology and therefore its contribution to biodiversity, this recognition of value is from a conservation perspective rather than a general perspective, indeed, something may be especially valued if it is of particular significance to conservation.)	Kites are seen as valuable in increasing the biodiversity of Britain and the complexity of its ecosystems. They were of particular interest to conservation in Britain, given that they were almost extinct there.
Economic	An entity is valued for its ability to be monetised and/ or the ability to measure its contribution to systems of capital.	Kites are valued as bio-capital and are valuable for the ecotourism which can be built around them (and monetised).
Existence	An entity is valued for the fact of its existence. An entity can hold existence value even if the person from whose perspective it is valuable has no engagement with it, or intention to engage with it in the future.	Kites' presence in Britain is valued, this value increased when the kite was facing extinction since its existence was threatened.
Functional/ ecological/ environmental	An entity plays a role in an ecosystem and is valued for its contribution to natural systems and processes.	Kites are valued for the role they play in Britain's ecosystems.

(Continued)

Table 6.1 (Continued)

Value	*Description*	*Example in relation to kites*
Instrumental/use	An entity has a utilitarian value and is valued for its practical purpose, as a means to an end (rather than as an end in itself).	As scavengers, kites were seen as valuable in keeping roads and streets clean.
Intrinsic	An entity has inherent value whereby it is valued for its own sake, as an end in itself (rather than as a means to an end).	Kites are seen as having value in and of themselves, demonstrated by the desire to protect them when they were almost extinct in Britain and the related desire to restore their population.
Novelty	An entity is valued because it is new, unfamiliar, and/or unusual and therefore interesting.	Recently (until their reintroduction), kites have been an unfamiliar sight in Britain. Seeing them is therefore a new and interesting experience for people.
Option	An entity is valued for its potential, even if it is not immediately valuable at present.	Kites in Britain now have an option value as a potential donor population for reintroductions in other countries where populations are declining.
Relational	An entity is valued as a result of its connection with another entity.	Kites have acquired value as a companion species, with people inviting them into their gardens through feeding.
Scarcity	An entity is valued for its rarity. Often the value increases inversely proportional to the availability of the entity.	Kites were valued for their rarity, as charismatic wildlife, but also as specimens for taxidermists and egg collectors.
Socio-cultural	An entity is valued as a result of its meaning to human society.	Kites have acquired a cultural significance and therefore value, particularly in Wales, where they are the national bird.
Stake	An entity is valued because of resources that have been invested in it that cannot be recovered (sunk costs). Often, the value increases in proportion to the level of sunk costs.	Kites involved in the reintroduction programme had high stake value because so much had been invested in their translocation.

recently, the kite has gained another form of *economic value* as bio-capital, given the establishment of an ecotourism industry around kite feeding. The rise of ecotourism associated with kite feeding is partly related to their scarcity value, meaning that people are prepared to travel and pay to witness them feeding, and also to their *aesthetic value* and *novelty value*. According to human tastes, kites are considered an attractive, even charismatic bird and therefore worthy of viewing. This is compounded by the fact that kites are willing to feed in large numbers, creating a considerable spectacle for the human observer. Similarly, because kites have been an unfamiliar sight, they present a new and interesting experience for people, again making them worthy of viewing.

The rehabilitation of the kite, not only in terms of numbers but in terms of reputation, has seen it obtain a *sociocultural value*, even being adopted as the national bird of Wales (Welsh Government, 2020). Similarly, our affinity with the red kite has also seen it acquire *relational value*, given that we now value it as a companion species, inviting it into our gardens to feed. Lastly, kites in Britain could be considered to have an *option value* in that they could become (even more) valuable in the future. For example, since kites are declining in other parts of their range and since Britain is now home to over 10% of the world's kite population, British kites are being translocated to other countries to reinforce their populations. If kite declines continue elsewhere, the role of the British kite in acting as a donor population will make it even more valuable from a conservation perspective.

Human-kite relations and feeding kites

While the categorisation and valuation of other-than-human animals are mutually influential, it is equally the case that the way we categorise and value animals dictates our relationships with them. One of the clearest manifestations of human-kite relationships is via feeding; humans feed kites in many different ways with significant variation depending on the relationship and the situation (see Table 6.2 for an overview). Kites are fed as wildlife and/or as pests, and while at times it is possible to differentiate between the two, at other times it is not. This is the case with *unintentional feeding*, where human activity provides resources that kites exploit as food (e.g., roadkill, food waste or food provided for other animals). In such cases, the resource is available to all kites and their categorisation (as wildlife or pest) depends entirely on the perspective of the human witnessing the feeding. At other times, however, the differences in how people feed kites, including their intent, are easy to identify. For example, when kites are classified as pests, humans engage in *necro-feeding*, feeding kites poisoned bait as a lethal form of control. While poisoning, or other persecution, of raptors is illegal in Britain under the Wildlife and Countryside Act (1981) necro-feeding is still conducted illicitly by those (often gamekeepers) who perceive kites as conflicting with human interests or values: 'there's a few places in England where people are still killing kites and it's all the same kind of issues that forced them to extinction in Scotland and England in the first place, so in particular issues on driven grouse moors' (Gareth). Indeed, Fred describes kites as 'extraordinarily easy to kill' because their nature as scavengers means that they are likely to take poisoned bait. Participant 19's use of the phrase 'extraordinarily easy' illustrates the casual and wholesale violence towards kites by those who consider them pests.

Table 6.2 Types of human feeding of wild animals with examples in relation to red kites

Feeding	*Description*	*Example in relation to kites*
Captive	Provisioning of food to wild animals in captivity. This feeding is usually the sole or main source of food for the animals in question and should therefore, ideally, meet all the animals' physical, nutritional and behavioural needs with respect to diet and feeding. The animals concerned may be in captivity for a number of reasons including, but not limited to, breeding programmes, translocation or release programmes, rehabilitation or conservation promotion via public entertainment/education.	Fledgling kites in captivity during translocation and prior to release at recipient sites in England and Scotland.
Diversionary	Provisioning of food to direct wild animals away from one area or behaviour (usually of potential or actual human–wildlife conflict) towards another (usually with lower potential for human–wildlife conflict). Diversionary feeding can include, but is by no means limited to, aversive conditioning and is a means of mitigating human–wildlife conflict, which can avoid lethal management methods (Kubasiewicz et al., 2016).	Feeding kites post-release anchored them to their release sites, reducing the likelihood of them dispersing into areas of potential conflict.
Necro	Provisioning of food in the form of poisoned bait or as a lure into a kill trap as a form of lethal control of target species.	Kites are poisoned (commonly by carcasses laced with poison) by those who view them as pests (e.g., gamekeepers).
Opportunistic	Provisioning of food in public or private spaces to facilitate close interaction with wild animals (Dubois and Fraser, 2013).	Kites are fed in people's private gardens, enabling people to see the kites and facilitating interactions with them.
Supplementary	Provisioning of food to complement wild animals' foraging or hunting behaviour to improve survival or reproduction (Dubois and Fraser, 2013; Kubasiewicz et al., 2016).	Kites were given supplementary food post-release in order to support their survival.
Tourism	Provisioning of food to make wild 'animals predictably and reliably viewable' in order to support wildlife tourism (i.e., 'non-consumptive interactions with wild animals') (Dubois and Fraser, 2013).	Kites are fed at feeding stations.

(Continued)

Table 6.2 (Continued)

Feeding	*Description*	*Example in relation to kites*
Unintentional	Behaviour which creates a food source for wildlife, even if the behaviour did not deliberately seek to do so, for example, feeding companion animals, planting certain vegetation and discarding or managing waste in a way that renders it available to wildlife. This feeding may be accidental, involuntary, unplanned and even unconscious.	Roadkill, food waste, and food provided for other animals provide food for kites to scavenge on.

Note: NB the animals involved may be classified as 'wildlife' and/or 'pest' depending on the type of feeding.

Source: Adapted and expanded from Thomas, 2025.

Alternatively, when kites are classified as charismatic, native wildlife, they are fed in very different ways, which, in Britain, largely relates to their reintroduction. Since the reintroduction of kites to Britain began with the translocation of fledglings, there was a period of *captive feeding* while the kites were transported and housed prior to release. This captive feeding required humans to provide food for the kites that met their nutritional, behavioural and physiological needs while also maintaining a distant relationship with the kites; this was intended to ensure that kites would be fit for release and could survive independently post-release, without having formed an association between people and food. One way of doing this was through specially designed feeding hatches in the rearing cages, as explained by Harry, a conservation consultant: 'we fed them through a hatch where just an arm appeared and put some food in'. This is notable since feeding often fosters some level of intimacy or even companionship through interaction, hence the deliberate steps having been taken to avoid that occurring here, something which is very different from the tourism and opportunistic feeding discussed later.

Following the kites' release, the feeding continued via *supplementary feeding*. Since on release, kites gain agency over their own feeding, humans are no longer solely responsible for meeting their nutritional, behavioural and physiological needs. Supplementary feeding is therefore intended to complement the kites' own hunting and foraging and to support their transition from captivity to the wild (i.e., to improve the likelihood that they would survive and reproduce successfully), as explained by Charlie, a policy expert:

> you're putting the food out to prevent excessive wandering initially but then also it's important to try and reduce that overwinter mortality, one of the things that is going to increase overwinter mortality with juvenile birds is starvation just because you have some birds that aren't quite

experienced enough yet, aren't able to provide enough food for themselves, and will die of starvation or that starvation is going to exacerbate some other disease issue that might otherwise not have been lethal.

Supplementary feeding demonstrates the care being extended to kites (in supporting their survival) but, equally, demonstrates the violence being enacted upon them through (partial) withdrawal of responsibility for their wellbeing and the risk of starvation that they are knowingly exposed to. The comment from Charlie also draws attention to *diversionary feeding* of kites, with the food preventing 'excessive wandering' and thus anchoring kites to their release site. Given the at times fraught nature of human-kite relationships, diversionary feeding can encourage kites not to disperse from release sites towards areas of potential human-kite conflict, such as the grouse moors discussed above (Dubois and Fraser, 2013; Kubasiewicz et al., 2016).

Supplementary feeding of kites was, arguably, required for only a very short time since the kites exhibited a high degree of self-sufficiency post-release, as was explained by Fred, a conservation consultant: 'they were pretty good at finding their own food after about two months'. Interestingly, however, feeding continued in proximity to several of the release sites, ostensibly as supplementary feeding but morphing, to a greater or lesser extent, into *tourism feeding*, which still occurs at 'kite feeding stations'. At the time the research, on which this book is based, was conducted there were three kite feeding stations in Scotland (Argaty Red Kites in Perthshire, Bellymack Hill Farm in Dumfries and Galloway and Tollie Kites in the Highlands) and three in Wales (Bwylch Nant yr Arian in Ceredigion, Gigrin Farm in Powys and Red Kites Wales in Carmarthenshire). All the sites took different approaches to red kite feeding, and while all claimed to be providing supplementary feeding, they were all also, indubitably, engaging in tourism feeding. Red kite feeding stations provide food at set times to make the kites 'predictably and reliably viewable' (Dubois and Fraser, 2013), with all the feeding stations advertising feeding times on their websites (Argaty Red Kites, Bellymack Hill Farm, Bwylch Nant yr Arian, Gigrin Farm, Red Kites Wales and Tollie Kites).

Kites are indeed predictably and reliably viewable at these feeding stations, gathering at the daily feeding time, sometimes in their hundreds, to take advantage of the food provided. While they are, to a large extent, scavengers, kites feed 'on the wing', swooping down to grasp a piece of food and flying away to eat it while airborne. This feeding behaviour, especially when performed *en masse*, creates a spectacular display and thus draws considerable numbers of visitors prepared to pay to watch this whirling dervish: all of the red kites feeding stations charge an entrance fee with the exception of Bwylch Nant yr Arian (Tollie Red Kites used to ask for a donation but is now permanently closed). While the visitor fee underlines the tourism element of kite feeding stations, all the feeding stations stressed the conservation aspect of their work, both in terms of the supplementary feeding of the kites and in engaging with visitors. They argued that the spectacle of kite feeding gave visitors the opportunity to appreciate the story of the kites' decline and restoration and to engage with conservation more generally. Feeding stations also stressed that the food they provided was supplementary and formed only a proportion of the kites' overall diet. For example Sophia a conservationist, said 'I don't overfeed, they've got to hunt for themselves', Gareth, another conservationist, said 'we give them a small supplementary feed to top up what they find in the wild' and Ella, also a conservationist, said 'we only feed a small amount each feeding day and the kites in no way rely on this, it is more like an extra snack to them'. The claim of supplementary feeding is, however, somewhat tenuous since sites take an unscientific approach to feeding (Sophia measured how much was fed in terms of 'half or three quarters of a bucket' while Gareth spoke of 'a rabbit's worth' of food) and have no way of regulating how much food an individual kite gets. This inability to regulate individual feeding was shown to have extremely serious consequences for kites, with at least one case where the feeding had led to the death of a kite, as described by James, a conservationist:

> it stood there filling its crop, it tried to fly across the lake and a gust of wind pushed it into the lake and it couldn't seem to get up and basically just the weight of the food in the crop weighed it down so much that it just couldn't get out and it unfortunately drowned.

Comments such as this, which reveal harm rather than benefit to individual kites, raise questions over whose benefit the kites are being fed for – their own or the spectators (or indeed the organisations running the feeding stations). While it is argued that the feeding is supplementary and supports kites, it can also be argued that it is tourism and benefits humans while actually harming kites. Kites are subject to (potential) harm in the interests of human entertainment, as the drowning of the kite illustrates.

The harms to other-than-human animals versus the benefits to humans are also evident in *opportunistic feeding*. Opportunistic feeding is evident where people living in proximity to kite reintroduction sites feed kites in their gardens, providing food as a way of seeing and interacting with them, as explained by James:

> we sometimes see up to twenty five kites in the village coming down and feeding in the back garden, and there's another one just across the road from me and that lady she probably gets ten or a dozen kites coming in, and you see that in quite a few other areas, once one or two kites find it and if you start to regularly feed them, then they'll keep coming back and then others will catch on the idea and then you get more and more birds coming in.

Charlie, a policy expert, recognised that this practice 'brings a huge deal of pleasure to people' but he considered it to be of questionable benefit to the kites and that it may even be harmful:

> where kites are getting a big proportion of their diet from being fed in back gardens, it would be great if people were feeding kites nutritious 'kite food', but they're not, they're feeding them sausages and bread and so I think there were issues to do with productivity, possibly lack of calcium, lack of other nutrients that the birds weren't securing because a large proportion of their diet is just junk food essentially.

Moreover, while the people who engage in kite feeding presumably view kites as valuable wildlife to be supported and encouraged, it is equally plausible that others living in the vicinity have a very different view of the kites, potentially leading to human-wildlife conflict.

Care, control and (renewed) co-existence

Care, control and violence

The different ways kites are fed offer very clear illustrations of the different ways humans care *for* and *about* kites, and wildlife more generally. Very broadly speaking, humans care *for* wildlife, which they value (with this value increasing if the wildlife is native, rare and charismatic), and care *about* pests, which they do not value. Thus, kites are generally cared *for* when they are categorised and valued as wildlife and cared *about* when they are categorised and not valued as pests. Care inescapably involves control, which can be benign but can also be coercive or even violent. Therefore, logics of care may involve seemingly contradictory elements of concern and consideration, coupled with coercion, violence and even harm (van Dooren's (2014) 'violent care'), particularly when viewed at different levels (e.g., individuals versus species). Care for and care about both entail control, and this can be illuminated by examining kite feeding practices more closely.

When categorised and valued as wildlife, kites are cared for, hence the motivation to conserve them, especially when they become rare. It was the desire to conserve kites in Britain, in the face of their imminent extirpation, that led to their reintroduction. While the reintroduction programme was benevolent in intent, it was also highly interventionist, involving extremely high levels of human control of and therefore power over kites, with significantly diminished kite agency. As a result of this power imbalance, there was significant scope for coercion, violence and even harm to the kites involved. In the case of captivity and captive feeding during translocation and prior to release, humans could be said to have had total control over the kites, including what they ate. The curtailment of kite agency, the restriction of their liberty, and the removal of their ability to feed themselves gave humans great power and ultimate responsibility in caring for the kites. While there was undeniably care for the individual kites involved (not least because of their stake value) this was arguably dominated by care for the species as a whole which was used to 'justify' control over and violence towards individual kites in restricting their liberty and even endangering their lives through the reintroduction process as a whole, in the interests of conserving the species.

Even after their release, supplementary feeding exerted a form of control over the kites and limited their agency to some extent. Given that they are facultative scavengers, kites are highly likely to exploit the food resources provided by supplementary feeding. While strictly speaking, kites could exert their agency by *not* utilising the supplementary feeding, this is essentially a Hobson's choice since *foregoing* the supplementary feeding is not a desirable alternative from the kites' perspective, and kites may even be disadvantaged if they do not exploit supplementary food resources. On the other hand, if kites *do* exploit the food resource, a form of dependency can be established which exerts a level of control over them and restricts their agency with Noah, a conservation project manager, admitting that he could not stop feeding the kites suddenly but that 'it'd have to be done over quite a length of time, putting out less and less meat'. Furthermore, since supplementary feeding can anchor kites to the feeding site, this has a further effect in restricting their liberty (the same point could be said to apply to tourism feeding, as will be discussed).

Diversionary feeding is perhaps even more controlling than supplementary feeding since it deliberately seeks to govern kite movement. As with supplementary feeding, diversionary feeding stems from a logic of care and has benevolent motives, indeed, one of its goals is to avoid human-wildlife conflict and perhaps more violent interactions or forms of control. Nonetheless, diversionary feeding curtails kite agency, using food as a form of biocontrol as much as to sustain life, and essentially impinging on kite freedom[2]. Similarly, although of benign intent and although, at least ostensibly drawing on logics of care, it could be argued that both tourism and opportunistic feeding exert the same forces of control as supplementary and diversionary feeding, limiting kite agency, restricting their freedom and creating dependency. Furthermore, both types of feeding demonstrate other forms of violence. In the case of tourism feeding, kites are, arguably, exploited – fed as part of a tourism industry for commercial gain rather than for the benefit of the kites themselves, contrary to the narratives of care and benefits to the animals that accompany tourism feeding (Orams, 2002). There is, however, the argument that the care here could be viewed at the species level rather than at the level of the individuals involved. While the kites being fed may not benefit (and may in fact experience *dis*benefit), the species *as a whole* may benefit if people's perceptions of kites are altered by the experience. Indeed, this benefit could extend to other raptors and wildlife more generally, as was described by James, a conservationist:

> we don't need to feed the kites here they're doing very well in other areas where there's no kite feeding going on, we look at it as an education tool to talk about the story of how we nearly lost them and the successful conservation project to keep them and how they've flourished since then, so we look at it in that way.

This virtuous effect may not, however, be the outcome of opportunistic feeding. In this situation, humans are feeding kites for their own gratification, often with little knowledge of, or consideration for, what to feed the kites, or whether to feed them at all. As well as potentially doing harm as a result of the feeding itself, opportunistic feeding potentially exposes kites to further harm, both at an individual level and even at a species level. Although opportunistic feeding of kites is rewarding for those who instigate it, in that it facilitates close interaction with the kites (Dubois and Fraser, 2013), because it is essentially unplanned and unregulated, it may bring kites into proximity with humans who view them very differently from those who undertake the feeding. This can result in kites being considered a 'nuisance' (i.e., a pest) by those who are not involved in the feeding but are affected by it due to their proximity, with potentially violent, harmful reactions against individual kites or the species as a whole.

This categorisation of 'nuisance' can also be created by the kites' proclivity as a scavenger. This behaviour makes kites adept at utilising anthropogenic resources, whereby humans are essentially providing food for kites unintentionally. Unintentional feeding has perhaps even greater potential for human-kite conflict than opportunistic feeding since it involves kites exerting their unruly agency. This can be confronting to people, especially when, as in the case of kites in Britain, people are not used to co-existing with them, kites having been largely absent from Britain for so long. This is exacerbated by kites being present in high numbers and/or in urban areas, which can result in them being considered 'out of place', both of which contribute to their classification as pests.

Co-existence

Human-wildlife conflict arising from human-kite proximity, potentially exacerbated by human feeding of kites, raises interesting questions. From a conservation perspective, the reintroduction of kites to Britain has been highly successful, with kites having established a self-sustaining wild population and now being present in large numbers. For the same reason (i.e., high kite numbers), the reintroduction might be considered less successful from a social perspective since, in some quarters, kites are viewed as overabundant, with calls for their control. While this perspective may be at least partially attributable to kites' willingness to live at high densities, it is also related to social carrying capacity (as opposed to ecological carrying capacity) and human tolerance for other species. Social carrying capacity relates to how the number or density of a species, or their impacts, affects human experience. This experience is largely dependent on human perceptions and beliefs, which in turn affect what level of impact they will tolerate (Decker and Purdy, 1988). When kite numbers and population density rise, and when their behaviours are confronting to humans, they may exceed their social carrying capacity before they reach their ecological carrying capacity. This results in them being considered pests, as opposed to when their numbers are low and they are considered valuable wildlife. This was exemplified in relation to kites in Britain, or more specifically England, by Gareth, a conservationist, 'there are some places where there are issues, or perceived issues, of an expanding kite population and in Britain I think we're not particularly used to having to live alongside wildlife and kites push that a little bit and test people's patience'.

The salient point here is that the trajectory which attitudes to kite reintroductions have followed, from initial enthusiasm and low levels of concern to a loss of tolerance and high levels of concern, commensurate with rising numbers of kites, could offer insights into how attitudes to other reintroductions may evolve and therefore lessons in how harmonious human-wildlife coexistence might be achieved. If reintroduction projects are to be sustainable, (new) ways for humans and wildlife to coexist harmoniously need to be found. As is the case with kites, this is often related to human tolerance of the unruly agency of wildlife. In many countries, as in Britain, this tolerance has been considerably reduced as a result of the prolonged absence of species that significantly challenge human values or interests. Because of this loss of tolerance, and because humans often lack experience of living with the species involved in reintroduction programmes, scholars discuss the potential for *renewed coexistence* in which people *re*learn to live with reintroduced species and acquire a renewed tolerance for the unruly agency of other-than-human animals (Auster et al., 2022). This could involve an 'enlightened attitude' to wildlife, compared to attitudes of the past (Arts, Fischer and van der Wal, 2016) and a 'radical tolerance' in terms of biopolitical thinking (Campbell, 2006)

Biopolitics: kites as self-determining agents and expendable objects

In some ways, this enlightened attitude and radical tolerance is evident in the biopolitical modes we apply to kites. Of the four biopolitical modes identified in Chapter 5, two are particularly relevant to kites: those of self-determining agents and expendable objects.

Kites as self-determining agents

It is with regard to kites as self-determining agents that we see enlightened human attitudes towards kites and a radical tolerance for their unruly, other-than-human agency. This enlightened attitude is illustrated by the philosophy behind the kite reintroduction programme; people reclassified kites from 'pest' to 'wildlife' and recognised the moral imperative of restoring a species both for its own sake and because humans were responsible for its decline (Arts, Fischer, and van der Wal, 2012; Crowley, Hinchliffe and McDonald, 2017). Such ecocentrism is an important element of biodiversity conservation and rewilding – reintroducing species in recognition of their intrinsic value rather than because of the roles they might play within an ecosystem or on behalf of humans (Taylor et al., 2020).[3]

It is not only in the moral rationale behind the reintroduction programme where a change in attitude is evident; although the initial stages of the kite reintroduction programme involved very high levels of human intervention, once the kites were released they were able to act according to their own agency and had the ability to express this fully (notwithstanding the arguments outlined above that some forms of feeding constrain their agency to some extent), and were able to act as self-determining agents. In acknowledging kites as self-determining agents, humans recognise kite agency at the species level and value kites as a species rather than as individuals. As a result, while kites are recognised as having intrinsic worth, because they are valued at a species level, the deaths of individual kites are not viewed as significant, while the loss of the species would be. The reintroduction programme illustrates this very clearly. Kite chicks were given considerable care and attention in captivity, but faced a significant risk of death post-release. While this was considered undesirable, and while care was taken to prevent deaths if possible, it was also considered an acceptable and indeed necessary part of the reintroduction process; the interests of individuals were secondary to the interests of the species as a whole. As soon as kites were released, a let-live, let-die logic was applied to them, with humans not intervening in kite life or death.

Kites as expendable objects

Interestingly, the other key biopolitical mode assigned to kites in Britain is that of expendable objects. This mode is at the opposite end of the spectrum from that of self-determining agents in terms of human interference in kite lives and deaths, human tolerance for kite agency, and, therefore, the ability of kites to express their agency freely without human intervention. In this mode, individual kites, and even the kite as a species, are seen as dispensable and the death of kites is seen as insignificant. While, in theory, the logic of let die still applies, this mode can see significant intervention, or calls for significant intervention, into kites' lives to the point that a *make* die logic may emerge. This becomes evident where kites are seen to conflict with human interests or values, and although it is particularly prevalent in areas where kite numbers are high, it is also present in areas where kite numbers are lower, but the narrative of conflict with human interests is very strong, as with game shooting, for example.

In areas with high kite numbers, there can be a loss of tolerance for their unruly agency as social carrying capacity is exceeded, coupled with calls for kite agency to be constrained. In areas of perceived conflict with human interests, kite agency is often constrained via lethal control through shooting or poisoning, despite this being illegal. In both situations, individual kites and the species as a whole are not seen as valuable, and their interests are perceived as being secondary to those of humans. This lack of human tolerance for other species is very clear here – humans are unwilling to accept the (relatively minor) impact of kites on their lives and instead call for kites to be controlled with significant, often fatal, consequences for the kites. This logic is enabled by the narrative that kites are expendable, that their lives are not valuable, and that the deaths of individuals, or even the species, are irrelevant.

Re-evaluating kites and co-existence

Human-kite history in Britain, from the Medieval period to the present time, offers fertile ground for exploring human–animal relations. Human relations with kites, and the interrelated ways that people categorise and value kites have changed, and continue to evolve, depending on time, place and socio-cultural factors. In relation to socio-cultural factors, the way kites are categorised and valued can depend on which sociocultural group people belong to (e.g., conservationists and gamekeepers) or on the level of impact from animal presence or behaviour that people are prepared to accept. Kites can be categorised as charismatic native wildlife or as pests depending on how they are valued or the extent to which they are seen to conflict with human interests or values. Kites can be valued for their intrinsic worth, their instrumental, functional or biological roles, or their aesthetic or novel qualities. They may also be valued for their existence and/or their scarcity. More prosaically, they may be valued in economic, acquisition, stake, or option terms. Lastly, they may be valued as a result of the socio-cultural or relational attributes which we assign to them.

The programme to reintroduce kites to Britain provides a 30-year snapshot of human kite relations, particularly as manifest through feeding. Humans engage in kite feeding in many ways. Sometimes this feeding is opportunistic or unintentional, but at other times it is a planned intervention for the benefit of the kites (captive, supplementary, or diversionary feeding), for the benefit of humans (tourism feeding), or to the detriment of kites (necro feeding). Feeding, as a form of care, whether *for* kites or *about* kites, inescapably involves an element of control and, often, elements of violence, even when the feeding is of benign intent. This control of kites can be viewed through a biopolitical lens, with humans identifying kites as self-determining agents and as expendable objects. The mode of kites as self-determining agents tends to be applied when kites and their behaviour are valued and tolerated. Conversely, the mode of kites as expendable objects tends to be applied when kites and their behaviour are not valued or tolerated. This is particularly problematic from the perspective of reintroduction programmes where animals are deliberately introduced to, certainly in the case of Britain, anthropogenic environments. In such cases, moving from a history of violence towards a harmonious (renewed) coexistence is paramount for the flourishing of the humans and other-than-human animals involved. In biopolitical terms, this involves a 'shift from a biopolitics as a *control over life* to a biopolitics of *living with*' (Lorimer and Driessen, 2013, emphasis added) – that is, of living with kites as self-determining agents. This will be considered further in Chapter 7, which examines the reintroduction of the wildcat (*Felis silvestris*) to Britain. As with kites, the case of the wildcat raises questions as to what humans consider valuable, especially in relation to conservation, and enables further exploration of how we categorise, control and coexist with other species.

Notes

1 While 'wild', 'wildlife', 'charismatic wildlife' and 'pest' are the important categories for the discussion of red kites in this chapter it is important to note that there are other categories which humans assign to animas, for example domestic, feral and hybrid, which will be discussed in relation to cats in Chapter 7.
2 It should be noted that diversionary feeding can exhibit still more violent care practices if, as can be the case, aversive conditioning it used (Kubasiewicz et al., 2016). It should also be noted, however, that this technique was not identified in relation to kites in Britain.
3 Anthropocentric rather than ecocentric motives for species reintroductions or involving other-than-human animals in rewilding projects are, however, evident in other biopolitical modes, particularly species as analogues and species as proxies.

References

Albert, C., Luque, G. M. and Courchamp, F. (2018) 'The twenty most charismatic species', *PLoS ONE*, 13(7), pp. 1–12.
Arluke, A. and Sanders, C. (1996) *Regarding Animals*. Philadelphia: Temple University Press.
Arts, K., Fischer, A. and van der Wal, R. (2012) 'Common stories of reintroduction: A discourse analysis of documents supporting animal reintroductions to Scotland', *Land Use Policy*, 29(4), pp. 911–920.
Arts, K., Fischer, A. and van der Wal, R. (2016) 'Boundaries of the wolf and the wild: A conceptual examination of the relationship between rewilding and animal reintroduction', *Restoration Ecology*, 24(1), pp. 27–34.
Auster, R. E., Barr, S. W. and Brazier, R. E. (2022) 'Renewed coexistence: Learning from steering group stakeholders on a beaver reintroduction project in England', *European Journal of Wildlife Research*, 68(1), pp. 1–22.
BBPC (2023) *Red Kite*. Available at: https://www.britishbirdofpreycentre.co.uk/conservation-projects/red-kite/ (Accessed: 21 June 2025).
Bird Life International (2020) *Species factsheet: Red Kite* Milvus milvus. Available at: https://datazone.birdlife.org/species/factsheet/red-kite-milvus-milvus (Accessed: 21 June 2025).
Campbell, T. (2006) '"Bios," immunity, life: The thought of Roberto Esposito', *Diacritics*, 36(2), pp. 2–22.

Carter, I. and Grice, P. (2006) *The Red Kite Reintroduction Programme in England*, Peterborough: English Nature.
Crowley, S. L., Hinchliffe, S. and McDonald, R. A. (2017) 'Nonhuman citizens on trial: The ecological politics of a beaver reintroduction', *Environment and Planning A*, 49(8), pp. 1846–1866.
Decker, D. J. and Purdy, K. G. (1988) 'Toward a concept of wildlife acceptance capacity in wildlife management', *Wildlife Society Bulletin*, *16*(1), pp. 53–57.
Dubois, S. and Fraser, D. (2013) 'A framework to evaluate wildlife feeding in research, wildlife management, tourism and recreation', *Animals*, 3(4), pp. 978–994.
Hird, M. (2009) *The Origins of Sociable Life: Evolution After Science Studies*. New York: Palgrave Macmillan.
IUCN (2008) *Global Reintroduction Perspectives*. Gland: IUCN. Available at: https://portals.iucn.org/library/node/47668
IUCN (2013) *Guidelines for Reintroductions and Other Conservation Translocations*. Gland: IUCN. Available at: https://portals.iucn.org/library/efiles/documents/2013009.pdf
JNCC (1996) *Review of Information, Policy and Legislation on Species Translocation*. Peterborough: JNCC. Available at: https://citeseerx.ist.psu.edu/document?repid=rep1&type=pdf&doi=1cb82ca11387e92ea76f5ca64d24e56c786d2669
Kubasiewicz, L.M. et al. (2016) 'Diversionary feeding: An effective management strategy for conservation conflict?', *Biodiversity and Conservation*, 25(1), pp. 1–22.
Lorimer, J. and Driessen, C. (2013) 'Bovine biopolitics and the promise of monsters in the rewilding of Heck cattle', *Geoforum*, 48, pp. 249–259.
Marino, F. et al. (in press) *Evaluating Motivations and Outcomes in Conservation Translocations*. People and Conservation Translocations, Exeter: University of Exeter.
O'Neill, J. (1992) 'The varieties of intrinsic value', *The Monist*, 75(2), pp. 119.
Orams, M. B. (2002) 'Feeding wildlife as a tourism attraction: A review of issues and impacts', *Tourism Management*, 23(3), pp. 281–293.
Overland, M. A. (1993) 'Poached eggs: England's collectors threaten species with an illegal hobby', *The Washington Post*, 13 August. Available at: https://www.washingtonpost.com/archive/lifestyle/1993/08/14/poached-eggs/77f41960-adac-4fa6-8c07-311eda5673be/
RSPB (2024) *Red Kites flying high - thanks to you!* Available at: https://www.rspb.org.uk/whats-happening/news/membership-thank-you-week-red-kites (Accessed: 21 June 2025).
Taylor, B., Chapron, G., Kopnina, H., Orlikowska, E., Gray, J. and Piccolo, J. J. (2020) 'The need for ecocentrism in biodiversity conservation', *Conservation Biology*, 34, pp. 1089–1096.
Thomas, V. (2025) 'The pros, cons and contrary consequences of conservation feeding: Anthropogenic feeding of the red kite (*Milvus milvus*) in Britain' in A. Mullan, R. Smallman, H. de Bondt, and J. Waterman (Eds.), *The Hand that Feeds: The Complex Relations of Human–Animal Feeding*. London: UCL Press. pp. 205–218.
van Dooren, T. (2014) *Flight Ways - Life and Loss at the Edge of Extinction, Nucleic Acids Research*. New York: Columbia University Press.
Welsh Government (2020) *The dramatic flight of the red kite celebrates Irish and Welsh cultural connections*. Available at: https://media.service.gov.wales/news/the-dramatic-flight-of-the-red-kite-celebrates-irish-and-welsh-cultural-connections (Accessed: 21 June 2025).
Wildlife and Countryside Act (1981) *The Wildlife and Countryside Act*. UK.
Williamson, T. (2013) *An Environmental History of Wildlife in England 1650–1950*. London: Bloomsbury.
Yalden, D. W. and Albarella, U. (2018) *The History of British Birds*. Oxford: Oxford University Press.

7 The wildness of cats

Caring for and about cats in Britain

Those interested in collective nouns may know that a group of domestic cats can be referred to as a 'clutter' or a 'cluster', while a group of wildcats is called a 'destruction' or a 'doubt'. These nouns, in their different ways, are particularly apposite in discussing the collection of domestic and wildcats in Britain. The situation is indeed cluttered, with 'untidy' assemblages of cats rather than clearly demarcated groups, which naturalists appreciate. Part of the blurring of cat categories is due to interbreeding and resultant introgression between domestic and wildcats, which has been referred to pejoratively as a 'cluster fuck' (Palmer and Thomas, 2023). The hybridisation of domestic and wild cats casts doubt both over the identity of individual cats and over the future of wildcats in Britain. Conservationists are concerned that it may lead to the 'genetic swamping', and therefore destruction of the wildcat population, indeed, wildcats in Britain have been classified as functionally extinct (Todesco et al., 2016). This chapter explores these issues, it examines the 'clutter' of domestic and wildcats in Britain and discusses attempts to categorise cats. In particular, the chapter examines attempts to segregate domestic and wildcats via efforts to definitively identify wildcats in contradistinction to domestic/wildcat hybrids.

The categories cats are assigned to dictate, and are dictated by, how people view 'cats', in the broad sense of the term. The interplay between cats being categorised as a *result of* the way they are perceived and valued, and their categorisation *influencing* the way they are perceived and valued, is explored by interrogating wildcat conservation. This conservation entails care for and about wildcats but also care for and about domestic cats and, indeed, care for and about domestic and wildcats in all their various permutations: feral, hybrid and captive. One of the ways this care is manifest is via attempts to control cats, both through categorisation itself and, in more practical terms, through feeding. The human-cat relationships which result from this feeding are discussed, and this leads on to a broader discussion of biopolitics and the governance of the life and death of Britain's cats.

DOI: 10.4324/9781003456490-8

A clutter of cats

The European wildcat (*Felis silvestris*) is native to Britain and other parts of Europe, and, although it is not of conservation concern in other parts of its range, it is considered critically endangered or even functionally extinct in Britain (Todesco et al., 2016). Indeed, wildcats have been entirely extirpated from England and Wales, and only a remnant population remains in Scotland. This extirpation and near total extinction is due to centuries of persecution, which substantially reduced wildcat numbers and which is now compounded by habitat loss and, more significantly, interactions with domestic cats (*Felis catus*).

The domestic cat is descended not from the European wildcat but from the African wildcat (*Felis lybica*) and is therefore genetically distinct from the European species. The two species are, however, interfertile and produce fertile offspring. The resultant hybrids are therefore capable of backcrossing with either domestic or wildcats, leading to introgression of the wildcat population. The level of introgression in the wildcat population in Scotland has reached such an extent that it is now considered a hybrid swarm, and it is no longer possible to genetically distinguish between generations of hybrids. The level of hybridisation was a determining factor in classifying the wildcat as functionally extinct, meaning that the wild population is no longer viable and is unable to recover without intervention (Breitenmoser et al., 2019). While hybridisation is now considered the greatest threat to the wildcat in Scotland, it was, initially, as much a symptom as a cause of the wildcat's decline: where robust wildcat populations exist, hybridisation with domestic cats does not pose a significant threat to wildcats. This is because, where wildcat populations are high, wildcats are able to locate other wildcats to mate with, thereby reducing the risk of interbreeding with domestic cats. Even if interbreeding does occur, robust wildcat populations can either avoid or withstand backcrossing.

In Scotland, by contrast, the wildcat population has fallen sufficiently low that wildcats are more likely to encounter, and therefore mate with, a domestic cat than another wildcat. The situation is complicated by the circumstances surrounding domestic cats, which fall into three broad but by no means easily circumscribed categories: pet, stray and feral. Pet cats are those which have a close relationship with humans and for which humans accept responsibility (Palmer and Thomas, 2023). Stray cats are those for which, for whatever reason, the human-animal bond has been disrupted or disestablished: these cats are, however, still closely associated with humans and remain socialised to people and human society (Palmer and Thomas, 2023). Indeed, to a large extent they remain dependent on humans, particularly for food, whether this is provided directly (via deliberate feeding) or indirectly by cats utilising food resources made available to them via human activities, albeit unintentionally (e.g., refuse, food provided for other animals or preying on other synanthropes, e.g., rodents and birds).

Feral cats on the other hand are those for whom a relationship with humans never existed on an individual level (although they are descended from cats which had a relationship with humans in the past since they are and remain a domestic species): they are unsocialised to humans and, theoretically at least, live entirely independently from human society (Hill et al., 2022). The latter point is, however, important since, while, strictly speaking, feral cats live independently from humans, there can be degrees of ferality, with feral cats experiencing greater or lesser degrees of contact with human society (Palmer and Thomas, 2023). Moreover, ferality, and indeed wildness, is further complicated by the Anthropocene, which 'disrupts any lingering notion that we can think of the environmental and social realms as separate and separable' (Head, 2015, p. 318). In other words, while we may think of feral and wildcats living separately from human societies, given the far-reaching influence of humans, it is essentially impossible for any animal to live independently from the effects of human activity.

These categories (wild, hybrid, domestic, pet, stray and feral) are, theoretically, distinct and yet, self-evidently, the existence of some troubles the clarity of others. For example, feral and hybrid cats challenge the notion of wild and domestic being binary states, and it being possible to make a clear distinction between 'wild' and 'domestic' cats. Furthermore, while pet, stray and feral can be defined on paper, in practice, it can be very difficult to determine which category a cat belongs to (Palmer and Thomas, 2023). Despite these difficulties, categories and categorisation remain a very important part of wildcat conservation and cat management in Britain. The category a cat is placed in affects, and if affected by, the way it is treated, often with significant consequences.

Categorising of cats

Conservation very much depends on categories in order to determine what to conserve, prioritising the wild, the native and the rare over the domestic, the non-native and the abundant (Milton, 2000). Maintaining distinct boundaries between categories is thus a crucial element of conservation since, without the ability to differentiate between categories, it becomes impossible to know what to conserve (Milton, 2000). As a result, the intermingling of the categories of wild and domestic, which hybrid (and feral) cats represent, creates significant questions for conservation.

Conservation has traditionally valorised the wild and denigrated the domestic. Wildness has positive connotations with associations of nobility and romanticism (Scasta, 2019) while the domestic has been seen as diminished, mutated and inferior (Lorimer and Driessen, 2016). Furthermore, hybridisation between wild and domestic species is looked upon unfavourably by conservation on the grounds that the wild gene pool is polluted by domestic genes (von Essen, 2017). Interestingly, the human element is part of the issue here: because domestication is the result of human intervention in animal breeding, domestic species are viewed as the result of artificial rather than natural selection and therefore as human artefacts. By extension, the existence of domestic species and, as a result, their availability to hybridise with 'natural animals' (Wagner et al., 2010) or wild species, means that such hybridisation can essentially be viewed as a human intervention in nature rather than as a natural process. By contrast, where 'natural' hybridisation (i.e., hybridisation between two interfertile wild species) occurs, it is not only accepted but is recognised as producing new species.[1] Pertinent to the matter at hand, 'natural hybridisation' between European and African wildcats is recognised as occurring where their ranges converge (Ottoni et al., 2017). This is significant because it makes differentiating between 'pure' and 'hybrid' wildcats in Britain even more difficult since it is impossible to determine whether admixtures of European and African wildcat genes are the result of recent hybridisation between European wildcats and domestic cats or the result of natural hybridisation between European and African wildcats in the past. Despite this difficulty, and regardless of the arguably inconsistent attitude towards hybridisation, which accepts and even appreciates the new species developed by natural hybridisation while dismissing the hybrids of domestic and wild species, the overarching policy of wildcat conservation in Britain is to devalue domestic wildcat hybrids. As a result, a considerable amount of effort goes into attempts to distinguish between hybrids, which are not valued, and 'pure' wildcats (i.e., those which are not hybridised (Wagner et al., 2010) or at least have a very low proportion of domestic genes), which are highly valued. This process is, however, extremely difficult and relies on genetic testing and pelage scoring, both of which have limitations.

Natural hybridisation between European and African wildcats in the past means that there may be no wildcats in Britain in which genetic tests would not reveal at least some proportion of *Felis lybica* genes, with no way of determining whether this is the result of ancient or recent hybridisation. Coupled with this is the fact that genetic testing is expensive and requires cats to be caught in order to obtain samples for testing. The alternative method of pelage scoring, while technically possible in the field, still relies on good visualisation of cats (either in person or via photographs) in order for a thorough assessment to be made. Furthermore, due to the level of introgression, pelage scoring is no longer as reliable a means of distinguishing between hybrids and pure wildcats as it once was. For example, a cat may rate highly on pelage scoring, suggesting that it is 'pure' or has low levels of domestic cat genes, but a genetic test can indicate the contrary, as explained by Jacob, a researcher:

> the correlation between the two [genetics and pelage] is very good during most of the 20th century, and you get to the end of the 20th century, beginning of the 21st century, and the whole thing breaks down because the association between the genetics of the coat pattern and the genetics generally across the whole genome all just become so intermingled that they're not associated anymore.

Conservationists have attempted to reconcile this difficulty by classifying a cat as a wildcat if it has a score of greater than 75% on genetic testing. This number is, however, somewhat 'arbitrary', as explained by Tom, a conservation project manager:

> there is a genetic test and a pelt test to decide whether this is actually a wildcat or not, and there is an *arbitrary line* in the sand to decide 'that one is' and 'that one's not', because the reality is, they do hybridise with domestic cats, and have done for however long there have been domestic cats, and they do in Europe, so they're all a bit hybridy, but then they are two very closely related species that can interbreed, so there's always going to be a level of hybridisation, and there is an *arbitrary* 'this is a wild cat' and 'this one isn't' which does make it tricky, a lot of it is down to human interpretation and human definitions (emphasis added).

Another driving factor behind these attempts to determine whether cats are pure wildcats or hybrids is the concern, as explained by Ava, a conservation project manager, that hybridisation will result in 'genetic dilution to the point that there's no longer any detectable wild cat genome'. This situation is largely attributable to the fact that there are so few wildcats left in Britain (to the point that they are considered functionally extinct) as compared to the very high numbers of domestic cats. The population of owned domestic cats in the UK is estimated to be 10.8 million (PDSA, 2024). Meanwhile, there are an estimated 250,000 urban strays (McDonald and Skillings, 2021) and 1.2 million feral cats (Yamaguchi et al., 2015). This situation means that hybridisation with domestic cats 'is considered the biggest threat to the survival of wildcats in Scotland' (Saving Wildcats, 2023a). The fact that the wildcat population is 'diminished to barely reproducible numbers' while the domestic cat flourishes at 'inflated' levels (Asdal et al., 2017) lends weight to the argument that the final stage of domestication is the extirpation of wild counterparts (Zeuner, 1963). As a result, conservationists care deeply for the wildcat's future and care deeply about the domestic cat's impact on the wildcat. Evidently, however, the ability of conservationists to classify cats becomes extremely important since it affects whether cats are cared *for* or *about*.

Care and control of cats

Wildcats' nativeness, wildness and rarity have made them an object of concern for conservationists, leading to efforts to restore their population, which involves high levels of care for the wildcats involved. With respect to feral and hybrid cats, their non-nativeness, domesticatedness (or taint of domestication) and (over)abundance[2] coupled with their impacts on the wildcat, also make them an object of concern for conservationists. Rather than leading to care *for* feral and hybrid cats, however, this leads to care *about* them. As a result of these different logics of care, conservationists treat wildcats and feral and hybrid cats very differently. One of the clearest manifestations of these different forms of care, and different forms of treatment, is through feeding – this is discussed in the following sections in relation to (i) wildcats in captive breeding programmes, (ii) wildcats for release, (iii) wildcats post release and (iv) feral and hybrid cats. In each case, feeding is a form of care but can, nonetheless, be violent as well as nurturing and even when it is nurturing, can still involve elements of coercion and control.

Efforts to restore the wildcat in Britain centre on reinforcing the population in Scotland and reintroducing wildcats to England and Wales. This strategy is dependent on captive breeding programmes to produce wildcats for release. In turn, captive breeding programmes depend on intimate human-wildcat relationships, with humans controlling all aspects of wildcat breeding and feeding. The way wildcats are fed can vary significantly depending on the role they play in the programme (for breeding or for release), the stage of the programme (pre or post release), and even the programme itself (wildcats are being bred at several sites in England, Scotland and Wales, and different sites take approaches to feeding). Wildcat restoration also depends on the management of feral and hybrid cats. Again, this management fosters intimate human–cat relationships, with humans intervening in cat breeding and feeding. Care, control and wildness are evident in all cases and influence, and are influenced by, feeding.

Wildcats

In captive breeding programmes

Wildcats in captive breeding programmes have their feeding and breeding entirely controlled by humans. Such human intervention is usually associated with domestication and, although it stems from a logic of care, is highly controlling and even coercive. Wildcats are selected for inclusion in or exclusion from breeding programmes, removing their (wild) agency over reproduction. Indeed, even wildcats selected for breeding have no agency in their choice of mate since this is also controlled by human decisions regarding which cats 'should' breed with which, rendering wildcat breeding in captivity a case of artificial rather than natural selection. Although humans are breeding wildcats to be, or become, wild, artificial selection is usually associated with domestication as opposed to wildness.

Wildcats in captivity are also entirely dependent on humans for their food: their ability to hunt is removed and their agency with regard to feeding is constrained, even to the point that when cats cache food for consumption later, their keepers are inclined to remove it because of their obligations to provide fresh food (Rose, conservation practitioner). Furthermore, food is an important part of human-wildcat relationships, and it is essential for the success of breeding programmes that these relationships are positive; wildcats in captivity are more likely to breed successfully if they are not stressed, and a trusting relationship with their human keepers, facilitated through feeding, is very much part of this. Thus, while to some extent efforts are made to maintain the 'wildness' of wildcats, on another level this is deliberately compromised for the benefit of the breeding programme by creating a bond between keepers and cats. This was explained in detail by Archie, a conservation practitioner:

> they're not fed, completely, quietly, quietly, there is some form of relationship with the keeper, because what Marianne [Hartmann, an authority on wildcat breeding for release] found is that the breeding success of wildcats is down to their stress levels and the relationship with the keeper is a big part of that. If you go in there completely quietly, creeping around trying not to disturb things, to the wildcat that just says you're up to something and they're trying to sneak up on me and I'm really unsure and this is making me really uneasy. If you go in there talking to the wildcat, ensuring they can always hear you, there's always sound, and even blinking back to the wildcat as you would with a domestic cat, they will blink back, they will respond, they understand that this person is calm, they're not a threat, they're providing food, so that's all good and they're much more relaxed and as a result have much bigger litter sizes.

Keepers are, however, cognisant of the tension between the need to foster positive human-cat relationships and the need to protect a cat's wildness, and the delicate balance that this entails as Isla, another conservation practitioner, explained: 'we don't want them to have unnatural behaviours with us but when they're in the breeding centre it's important that they're comfortable with us, we don't want them to be domestic cats but they need to be comfortable with us, which they are'. Feeding wildcats in captive breeding programmes is therefore an intricate process which must achieve a delicate balance: there must be sufficient intimacy in the keeper-wildcat relationship for the wildcats to be relaxed enough to breed, but a sufficient distance to allow the wildcats to retain their 'wildness' and natural behaviours. The 'wildness' of the captive cats is, however, inescapably altered given the human intervention in and control of their breeding and feeding and the very existence of a human relationship with this highly anthropophobic species.

Pre release

The feeding of wildcats intended for release is very different from the feeding of wildcats for breeding. While it is still important that they are comfortable with their human keepers (e.g., for the purposes of health checks and procedures), the need is much lower than it is for the breeding cats and becomes lower still as cats get closer to being released. Thus, although feeding is an intimate form of care, the wildcat-keeper relationship is, by necessity and by design, more distant: keepers take care that the wildcats do not associate them with food and do not become habituated to humans. As part of this, keepers rely on the wildcats' innate anthropophobia to assist in maintaining a degree of distance in their relationship, retaining the 'wildness' of the cats, and ensuring that they are averse to humans on their release. Nonetheless, as explained by Arthur, a conservation project manager, wildcats in captivity are inevitably habituated to humans to some extent, thus compromising their wildness, although efforts are made to mitigate this during the pre-release stage:

> They've all come from captive collections so they will probably be reasonably habituated to human beings to greater or lesser extents. We're choosing out of preference the ones that have had less of that. Over the duration of the time that they're with us they will have a minimum of human contact so that *they are as wild as they possibly can be at the point when it comes to release them* (emphasis added).

Arthur alludes here to the question of the cats' wildness – whether wildcats in captivity are wild or not and whether wildness is a binary state or exists on a spectrum. This question will be discussed in more detail in the following section (The wildness of cats).

Perhaps an even more pressing issue than that of cats' habituation or aversion to humans relates to the need to prepare them for survival post release and, to a point, the two issues can be addressed in the same way, through 'remote' feeding. Remote feeding helps to avoid cats associating humans with food and also offers ways to simulate hunting, therefore eliciting behaviours that will be essential for the cats' survival post release. The paradox of simulating hunting, however, is that the simulation of the wild is highly contrived and exposes cats to more human artefacts. As Rose explained, hunting is simulated through the use of what are essentially domestic cat toys. Rose went on to acknowledge that wildcats interacting with cat toys is an inherently artificial situation and yet is stimulating, or attempting to stimulate, 'wild' behaviours:

> From an outsider looking in it just looks like they're playing with toys but what we need to do is develop those [hunting] skills, so feeding them in a way that doesn't necessarily replicate what they would do in the wild, because they wouldn't interact with a box in the wild, but it develops those [hunting] skills.

Simulating hunting, rather than feeding live prey, is necessary since the feeding of live prey is highly regulated in Britain. While in some cases derogations can and might (theoretically) be made in order to permit live feeding, none of the wildcat breeding centres in Britain have applied for such an exemption, saying either that they did not consider it essential for the health and wellbeing of the wildcats or that they considered it prohibitively difficult. This is fraught for two reasons: first, as with the feeding of wildcats in captive breeding programmes, the wildness and agency of the cats are compromised by their inability to hunt and feed without human intervention or constraint. Secondly, and with potentially more serious consequences, it is possible that the cats' preparedness for release is compromised due to their inability to acquire and develop hunting strategies and skills that will be essential for their survival post release. The denial of this opportunity is therefore, arguably, a violent form of care since it potentially jeopardises wildcat wellbeing. The two perspectives, that the inability to live feed may or may not be detrimental, were clearly illustrated in separate interviews. Theo, a conservation project manager, explained that:

> with live feeding in captivity, some argue that actually it only allows cats to practice a subset of behaviours associated with hunting because if you put a rabbit in an enclosure, the cat any point in time can really get that rabbit, that rabbit can't get out, it's not going anywhere, in order for the cat to catch the rabbit it needs to catch it in what is a small area. The rabbit might be able to get away for a short time, but at some point the cat will get it. It will catch the rabbit, it'll kill the rabbit and even if it doesn't kill the rabbit quickly or effectively it will get the rabbit again and it will do that. So, it's actually only a small selection of skills that the cat will really be developing, whereas there's a lot of skills like waiting, stalking, following scent trails, ambush behaviours, and so on that might not be developed just by catching and killing.

Meanwhile, Rose was open about the possible consequences of not being able to feed live prey, saying, 'maybe we'll release the first cats and none of them'll be able to eat anything and they'll all starve to death and then we'll know that actually maybe we need to feed them live prey'. Besides Rose's honesty, there is a pragmatism that borders on casual violence towards the cats: despite the intense care for the cats while in captivity, this care is withdrawn abruptly on release, with potentially fatal consequences for the cats. This level of care and the acceptance of the harm which the care, or its removal, entails is discussed in more detail in the following sections (Post release and Commanding cats).

Post release

The violence inherent in the release of captive animals to the wild, even when done with benevolent intent, is unmistakable – there is an abrupt, if not complete cessation, then considerable alteration in care and responsibility. Very different approaches are adopted by wildcat release centres across Britain, with participants discussing 'hard' versus 'soft' releases. Hard releases entail releasing wildcats at the release site without any preparation beyond that which has already taken place at the breeding centre and, usually, without providing any further support such as supplementary feeding. Amelia, a policy expert, suggested that wildcats are, in fact, unlikely to utilise resources such as supplementary food:

> one interesting thing about wildcats is that when they've been put out in the wild they don't seem to come back necessarily, they don't seem to be interested, they seem to just go off and feed themselves rather than taking food that's been put out for them. However, there is a slight caveat to that though because if they've been in this system where they've been partially soft released, then it seems to be a different matter, that they will then take food a bit more than they would have done, but generally they seem to ignore it once they're out in the wild situation. Once they've been able to go out they don't seem to come back for it.

Some of the release centres were planning to take the hard release approach, accepting the trade-off between the violence (arguably) involved in the withdrawal of care and the desire to relinquish control of the wildcats once they are restored to the wild, and once their wildness is restored to them. This approach was linked to a desire to avoid (further) impinging on the cats' wildness and can be viewed as related to ideas around 'intervening' in nature versus 'letting nature take its course'.

In stark contrast, other release centres were in favour of soft releases, which involve wildcats being housed in temporary pens at the release site for a short time. During this time, the wildcats remain captive, and the keepers remain responsible for meeting all their requirements in relation to feeding. On release, the pens are opened and the wildcats can leave according to their own agency. Their (erstwhile) keepers will, however, continue to provide food in the pens, which the wildcats can return to, again according to their own agency. Tom, a conservation project manager, argued that, while this might continue to constrain the cats' wildness, it gives them the best possible chance of survival post release:

> It's not just breed lots of animals, chuck them out and hope for the best, which is a very old-fashioned way of doing conservation releases, breed lots of animals, off they go, and oh god, fifty percent have died[3], which is not unusual, fifty percent mortality is not an unusual situation. Our view is we should be aiming for a far higher survival rate because you put a huge amount of investment into breeding these animals, it costs a lot of money, it takes a lot of time, they're very valuable genetic animals, we'll look after them the best we possibly can, we're basically getting our cats and going 'out you go, but don't worry guys, we've got some food for you, and lots of nest boxes around for you to shelter in' so it's a very different approach.

Supplementary feeding, such as described by Tom, maintains a distant human-wildcat relationship, but, while distant, and while designed to fade over time, the relationship may in fact endure for prolonged periods, possibly even spanning generations of wildcats and encompassing many conservationists, as he went on to explain:

> if the first generation are still quite dependent on being fed and they're still quite dependent on us putting out artificial nest boxes for them, that's fine. We're looking at generation after generation after generation, a long-term project. It goes counter to some old school conservationists who think once you've let it go, it's gone, you shouldn't monitor it, you should leave them alone. I don't agree.

Supplementary feeding thus seeks to mitigate the violence involved in (hard) releases and attempts to extend care beyond release to continue to support the cats. While caring, it is also controlling; however, potentially making the wildcats dependent on the food source and anchoring them to the release site, thus compromising their agency and indeed their wildness. Supplementary feeding is thus an excellent illustration of the tensions inherent in reintroduction programmes, in this case, the trade-off between impinging on the wildness of cats and increasing their chances of survival.

Feral and hybrid cats

Because feral and hybrid cats are not valued by conservationists[4] and because they are interfertile with wildcats (leading to 'pollution' (von Essen, 2017) of the wildcat gene pool), they have become objects of significant concern to conservation. As a result, there are concerted efforts to control feral and hybrid cats, which manifest as attempts to control their numbers through attempts to control their reproduction. It should be noted that while this is, ostensibly, an act of violence towards feral and hybrid cats in that it attempts to constrain their very existence and curtail their reproductive agency, it can, arguably, be seen as a form of care since it avoids other, more violent forms of control (i.e., culling). Violence is, nonetheless, evident throughout the process.

Control of feral and hybrid cat reproduction is achieved through 'trap, neuter, vaccinate, release' programmes[5]. As the name suggests, these are strategies wherein free-living cats are captured, sterilised and returned either to the area where they were caught or to another suitable location. Although the process is intended to be safe and humane, it is not risk-free and cats are liable to, at best, stress and at worst, injury as well as, *ipso facto*, violation of their liberty. Furthermore, since trapping is an indiscriminate process, captured cats must be assessed and categorised. As will be discussed in more detail later in this chapter categorisation is a form of violence and control in and of itself; the classification of cats as feral or hybrid consigns them to an undesirable category which is not afforded the same kind of care and concern as the more valued category of wild and, indeed, means that they will be subject to neutering. Although spaying and castration are considered routine procedures, and are statistically very safe, they are not entirely free of risk, and neither are they free of violence, violating as they do the cats' reproductive systems (c.f., Srinivasan, 2013). Finally, cats are 'released', which, while having positive connotations in that it restores their liberty, results in less surgical aftercare than would be provided to an owned cat. Perhaps even more significantly, the cats are consigned to 'neuter' lives (i.e., ones in which their ability to display natural behaviour and express their agency with respect to reproduction is removed) and feral or hybrid lives (*sensu* Johnston, 2021), condemned to the margins of our socio-ecological system, with humans neither accepting responsibility for them as domestic (a status which they retain despite their ferality and hybridity) nor affording them sovereignty (and value) as 'wild', despite their free-living status.

Despite the inherent violence of these programmes, there is also a significant element of care. In addition to the broader aspect of care demonstrated in that trap, neuter, vaccinate, release programmes avoid lethal control, the feeding involved entails considerable care and carefulness. Cats must be lured to and *in*to traps, which is achieved via food bait – a type of conservation feeding known as 'attractant feeding'. In order to accomplish this, the bait must first be sufficiently appealing to encourage the cat to enter the confined, unfamiliar and suspicious environment of a trap and, secondly, must be sufficiently inaccessible that the cat cannot seize it and partially or fully escape before the trap has had time to spring. This ensures that the cat is not unnecessarily stressed or even injured during the trapping process and also that it *is* trapped – if a cat escapes, it may become much more wary in the future and therefore much harder to trap.

Accomplishing this requires a high level of knowledge, care and attentiveness from conservationists to identify foods that will be sufficiently attractive to cats. This (procedural) knowledge may be highly specific to individual conservationists, as explained by Oliver, a conservation practitioner:

> everybody's got their own methods. Fish is quite good, tinned tuna and stuff like that, something that you can, depending on your trap, something that you can really squish into the bottom so they can't drag it away or pull it or anything, they have to go and actively try and get at it and they step on the treadle and trap. We're working on a trap at the minute where they're going to have to pull bait from the back and it'll close from the outside and it's super humane and very unlikely they'll get injured, it's the result of a lot of years of mistakes.

This quote is highly revealing about the care and violent care involved in conservation feeding practices. First, it highlights the care and knowledge involved in knowing and feeding cats, and knowing what food bait will attract a cat into a trap. Second, it reveals that there is more to attractant feeding than simply luring a cat into a trap. The food must be sufficiently attractive to entice the cat into the trap and should also be of a form that makes it difficult or impossible to take quickly without triggering the trap and being caught by it. Lastly, the final part of the quote reveals the violence inherent in this conservation practice of care and control. Conservationists may express concern for feral and hybrid lives, but the reality is that past and even present conservation practices involve threats to cat welfare. While, as Oliver's words relate, these threats are undesirable, they are also acceptable and accepted, albeit that conservationists and other stakeholders have been and are working to mitigate them.

The wildness of cats

As this chapter shows, the 'wildness' of cats is a matter of concern for conservationists and very much affects how they are valued and treated. In the case of the cats discussed here, their 'wildness' is primarily determined by their species (in very simple terms, *Felis sylvestris* are wild, *Felis catus* are domestic, although this is of course complicated by hybrid cats) but can also be influenced by the way cats live and the roles they play. Wildcats in captivity and free-living feral cats trouble the notion of any clear wild/domestic boundary, illustrating how wildness is affected by a cat's agency and its ability to live freely from human intervention. This is in-line with the definition of wildness as being 'the autonomy of the more-than-human world where events, such as animals moving about, plants growing and rocks falling occur largely because of their own internal self-expression that is independent of civilized forces' (Woods, 2005, p. 177). Here, wildness and agency are clearly connected, with wildness giving other-than-human actors the freedom to express their agency, independent of 'civilized forces'. Extrapolating from this, human interventions such as domestication and captivity curtail other-than-human agency and therefore wildness: human interventions such as these, even when they have ceased, can have lasting effects on cats' wildness.

Genetic wildness

Understandably enough, *Felis sylvestris*, as a wild species, is considered 'wild' while *Felis catus*, as a domestic species, is not. This distinction is, however, complicated by hybrids and the difficulty of distinguishing between 'pure' wildcats and those which are hybridised. Genetic testing is relied upon as a determinant of species, with genetic purity employed here as a proxy for wildness – the higher the proportion of *Felis sylvestris* genes (as opposed to *Felis catus* genes), the wilder a cat is. Given that, due to ancient and recent hybridisation, there are no 'pure' wildcats in Britain, conservationists are faced with the question of what degree of hybridisation is acceptable for a cat to be considered wild and included in the reintroduction programme, or allowed to continue living 'in the wild'. Since inclusion/exclusion is binary, conservationists must assign an equally binary state of 'wild'/'not wild' to cats and do so, as discussed earlier, according to the percentage of *Felis sylvestris* genes it has, with a score of 75% or greater allowing a cat to be classified as wild. Choosing a cut-off point on a scale from which to draw binary conclusions is, self-evidently, a matter of discretion, hence the recognition from Tom, the conservation manager quoted previously, that this the cut-off point is an 'arbitrary' one which rests on 'human interpretation and human definitions' which is directly related to (arguably equally arbitrary) classifications of species.

Functional wildness

Closely linked to the degree to which a cat is 'genetically wild' is its 'functional wildness' – the extent to which it plays the role of a wild animal in the wild. Ava, a conservation project manager, discussed the situation where a hybrid cat may not be considered *genetically* wild (i.e., it does not have sufficient *Felis Sylvestris* genes to be considered a wildcat) but may still *function* as a wildcat within an ecosystem:

> if a cat has got 90% wildcat DNA and it's living in the woods and it's eating voles is that legitimately still a wildcat or not? If they look like wildcats and behave like wildcats then they're wildcats. You know, what's the problem? You trap it and test it and find it's 98% or 61%, if they look the same and do the same thing, does that really matter?

Freya, a conservation practitioner, also recognised the reverse situation where a cat might be genetically wild but questions may be raised over its ability to perform as a wildcat due to its captive origins: 'a lot of people would say that we can't possibly release effective cats that will act as wildcats because they've been captive bred'. Essentially, being born in captivity and spending time in captivity may not only compromise an animals' wildness in terms of restricting its agency but also in limiting its ability to perform the role of a wildcat within an ecosystem.

What is in question here is what 'form' or wildness is more important, both in conservation terms and, more broadly, in terms of how humans value wildness. Do we consider it more important that a cat is genetically wild, even if it has been in captivity, and even if this limits its ability to function as a wildcat within an ecosystem, or do we consider it more important that a cat can function as a wildcat within an ecosystem, perhaps *because* it hasn't lived in captivity, and despite not having sufficient *Felis sylvestris* genes to be considered genetically wild.

Authentic wildness

A cat's origins (or perhaps its authenticity) is another major determinant of whether it is considered wild, with birth in captivity versus 'in the wild' being the crucial factor. As Lily, a researcher, put it, 'there's always those interesting questions about are those individuals that have been captive bred and then released truly wild'. This goes beyond Freya's comment, quoted above, about whether a captive-bred wildcat will be 'effective' in the wild, and questions the very *validity* of captive-bred wildcats *as wild*. Other research participants shared the view that captivity compromises a wildcat's wildness. Arthur, a conservation project manager, described wildcats as being 'as wild as they possibly can be at the point when it comes to release them'. Describing cats as 'as wild as they possibly can be' suggests not only that they are not 'fully' wild but also that wildness is a scale or a spectrum rather than a binary state. Other participants shared the view that wildcats bred in captivity, even once released, did not, perhaps *could* not, become fully wild with Ivy, a conservationist, suggesting that to *be wild*, a cat must be *born in the wild* 'it's the generations after the first lot have gone out, that's when I would see them as being wild, *if they're wild born*' (emphasis added). Tom, a conservation project manager, went even further to say that it would take *generations* of being born in the wild for cats to be 'truly wild': 'we're looking at generation after generation after generation, where actually maybe the third generation of kittens will be *utterly, truly* wild and have no supplemental feeding, not use nest boxes and just go off and do their own thing' (emphasis added). While Tom's view of the cats' wildness seems to hinge, at least partly, on the ongoing human intervention in their lives, and thus the reduction in their (wild) agency, it also appears to rest on the degree of separation from (being born in) captivity. This suggests that a 'hallmark' or 'taint' of captivity remains in captive-bred wildcats, perhaps in the way that the 'hallmark' of domestication remains in animals which become feral but are never able to regain their wildness (Palmer, 2016).

Thus, a wildcat in captivity may not be considered 'fully wild' due to the compromise of its agency, and this can extend beyond release so that a cat does not automatically gain 'truly wild' status the moment it is released. Indeed, a captive-born wild cat may never gain wild status, especially if it is dependent on ongoing human intervention rather than being self-sufficient. It may take one or many generations for wildness to be achieved (although again, this may be compromised if there is human intervention). Furthermore, extrapolating from the argument that the hallmark of domestication remains in feral animals, rendering them unable to ever (re)gain wildness (Palmer, 2016), it could be said that the authenticity of reintroduced wildcats is compromised by them having been captive bred since they exist as a result of 'artificial' rather than 'natural' processes (Katz, 1997). This argument would mean that reintroduced wildcats can never be seen as 'truly' wild because they can never escape their 'artificial' origins (Katz, 1997).

Blurring wildness and new wildness

What these categories (genetic, functional, authentic) of wildness demonstrate is that hybridity, captivity and even ferality blur the concept of wildness, intermingling it with domesticness and tameness. Participants used the terms 'blurred' and 'blurring' to acknowledge that the distinction between what is or is not wild is not necessarily clear. Ivy, a conservation practitioner, recognised this in relation to hybridity, saying, 'when you've got hybrids that look like wildcats, it makes it all a bit blurred'. Tom, a conservation project manager, went further, saying that *all* wildcats are a bit 'hybridy' due to ancient as well as recent hybridisation. Amelia, a policy expert, went further again to say that, because of the level of hybridisation, 'you can't really call what's out in the wild, wildcats particularly'.

Of interest is how these hybrid cats behave and function in the wild. They blur the boundaries of wild and domestic by bridging the two categories and occupying a liminal zone between them. Oliver, a conservation practitioner, described this zone as a 'grey area' with hybrid cats moving back and forth between populations of feral cats and wildcats, acting as a 'bridge between the two'. Ava, a conservation project manager, used the same concept of hybrids 'forming a bridge' between domestic and wildcats, but went even further to say that they were 'occupying an ecological niche in the middle of domestic cats and wildcats'. This is very much related to the functional wildness of cats and whether, in the Anthropocene, feral and hybrid cats might be as, or even better equipped, to fill new ecological niches and play a role in the novel ecosystems that are emerging. Ava expanded on this point saying that hybrids are 'probably more likely to occupy intermediate environments, they're likely to be less intimidated by people and more likely to use quite human dominated landscapes' and that as a result 'it looks a lot like hybrids are probably going to be much better at surviving out there than wildcats might be because we have hybrid landscapes, we have hybrid habitat out there and hybrids survive in it and wildcats don't.'

As a result, despite conservation conventionally valorising the wild and marginalising hybrids, hybrids may acquire value in the Anthropocene if their morphophysiology means that they are well suited to the 'hybrid landscapes' in which they find themselves and if they can contribute biological and functional diversity to an ecosystem where those traits are otherwise missing (Palmer et al., 2021). Such a shift would require a degree of pragmatism, but this is already in evidence in the acceptance of cats with 25% *Felis catus* genes as 'wildcats'. Moreover, the prevalence and strength of such attitudes may rise if the ongoing Anthropocene increases conservation pressure on species and habitats (Palmer et al., 2021). Until that time, however, hybrid and feral cats (and many more other-than-human animals involved in reintroductions and rewilding) continue to exist at or in the margins, within and between the boundaries of wildness and domestication, and this very much affects how they are governed.

Commanding cats

As with red kites, the biopolitical modes which are most relevant to the cats in this chapter are those of self-determining agents and expendable objects, although there is also something very interesting to be said in relation to cats as analogues. It should be stated at the outset that the categorisation of cats (as wild, domestic, hybrid, feral) is, in and of itself, 'a biopolitical technology par excellence' (see Braverman (2016) who discusses this in relation to threatened species lists, something which is highly relevant here since wildcats are categorised as critically endangered in Britain). Once categorised, cats are subject to further biopolitical control depending on the category to which they are assigned.

Self-determining agents

Cats which are categorised as wild and which are released into the wild can be considered self-determining agents from a biopolitical perspective. Post-captivity, wildcats acquire a much greater ability to express their agency than they have while in captivity. As with kites, however, this agency may be constrained by ongoing human intervention. In some cases (as discussed above), this intervention will be in the form of supplementary feeding. Other interventions which have not yet been discussed, and which are extremely interesting from a biopolitical perspective, are Global Positioning System (GPS) collaring and camera trapping.

Post release, wildcats may be subject to surveillance via GPS collars, which transmit their movements and location (Saving Wildcats, 2023b). In addition, trail cameras may be set up in the vicinity of release sites to further monitor wildcat activity, often using bait (i.e., attractant feeding) to lure the cats to the cameras. This intensive surveillance is part of the transition of the wildcat from captive to free and part of the shift in care from individual to species. While captive, wildcats are cared for as individuals, once released, the logic of care is applied at the species level. Wildcat lives are valued as part of functioning ecosystems, and while individual wildcat deaths are undesirable, they are acceptable within the wider scheme of reintroducing the species to the wild (indeed, as discussed earlier, participants acknowledged that mortality rates are likely to be very high). The abrupt withdrawal of care for individual cats is confronting for those involved, and potentially fatal for the cats, but is seen as a necessary trade-off in the short term in pursuit of establishing a flourishing, self-sustaining wild population in the long term.

While done with benign intent (Noddings (1999) might even call it benign neglect), the abrupt withdrawal of care is an act of violence for the cats involved – cats are let live but they are also let die, in theory, without human intervention to prevent this, despite the fact that the degree of surveillance could make such intervention possible. That being said, because individual deaths are highly undesirable (perhaps even more so than is usually associated with other-than-human animals to which this biopolitical mode is applied) the logics of care and governance (i.e., of non-intervention and of let-live, let-die) which usually apply to species as self-determining agents will be disrupted somewhat by supplementary feeding in some cases. On the other hand, it could also be argued that this is a pragmatic response since the survival of the species is so desirable (and tenuous) that efforts to avoid individual deaths are as much about care for the species as a whole as for individuals themselves. Nonetheless, as with kites, while done with benign intent, supplementary feeding impinges on the agency of wildcats and their ability to function as fully self-determining agents.

Expendable objects

In sharp contrast, cats which are classified as hybrid (having more than 25% *Felis catus* genes) or feral are considered expendable objects. The logic of care towards them applies at the individual and species levels; *Felis catus* as a species is not considered valuable by conservationists, and its members are considered dispensable in the interests of wildcat conservation. This is exactly the kind of 'species thinking' and 'ethical taxonomic categorisation' that makes members of certain species 'expendable forms of life in the service of other, more needy, beings' making them '"killable" in the name of the greater good of conservation' (van Dooren, 2014, p. 116–117). In Britain, conservation and public sensibilities towards killing cats prevents these cats from actually being killed (see Palmer and Thomas, 2023) but their violent enrolment in trap, neuter, vaccinate, release programmes illustrates that they are, nonetheless, essentially killable, or at the very least that their genes are killable, with the ultimate goal being to remove feral and hybrid cats (and their genes) from the landscape.

The violence toward these cats and the extreme biopolitical control to which they are subject[6] demonstrates clearly that their lives are not valued and that their deaths do not matter. Indeed, their deaths are not merely irrelevant; they may, in fact, be *desirable*, with the cats seen as dispensable. There is a sacrificial logic at play here where the agency and lives of these cats can be sacrificed in the interests of wildcats, with the logic of death leaning towards 'make die' as much as let die (as was also seen in relation to red kites).

Analogues

Despite being discussed above, as expendable, it is important to note that hybrid and feral cats can be different things to different people. While the dominant view in conservation currently is that hybrid and feral cats are essentially, 'pests' (even to the point of being 'invasive'), 'non-native', 'over abundant', 'tainted by domestication', and generally a threat to the valued wildcat, this view is not necessarily universal and may not necessarily be the dominant view for all time. There is an alternative view that hybrid cats, in particular, although also feral cats, could be, or have the potential to be, analogues for the 'pure' wildcat. This leads to perhaps the most interesting (potential) biopolitical mode in relation to cats, where hybrid and feral cats could be valued at the 'species' level for their ability to fill the ecological niche of the wildcat. In this mode, hybrid and feral cat lives would be valued, although more for the role they would play and the species they would replace than for their own sake. Furthermore, conservation efforts would no longer be focussed on attempts to eradicate hybrid and feral cats and, rather than the logic of *make* die, there would be a logic of *let* die, with death becoming accepted rather than desired. Indeed, there would be far less human intervention in hybrid and feral lives, and cats would have increased ability to express their agency, particularly their reproductive agency.

It must be acknowledged that a major shift in the conservation paradigm would be required for this biopolitical mode to come into play. Current conservation convention argues that feral and hybrid cats play different roles in ecosystems than wildcats, in terms of diet and preferred habitat. Tom, a conservation project manager, explained the difference in prey preferences: 'domestic cats eat lots of birds. Wildcats, they will eat birds, but actually they prefer rabbits, voles, mice, rats, they prefer rodents, and if they have a prey choice that tends to be what they go for.' Meanwhile, Alfie, a researcher, described the difference in habitat, while also alluding to diet:

> feral cats tend to be found much closer to human settlements, they tend to have a much higher proportion of human associated foods in their diets compared to a wildcat which tends to roam further, it's found more in woodland edges, has a purely wild diet and is inclined to separate itself from humans and find niches in undisturbed areas.

Jacob, another researcher, also acknowledged, however, that hybrid cats, particularly those which are 'more like wildcats', act 'virtually like wildcats' and that there is 'a great deal of overlap in diet between the hybrid population and the wild population'. Furthermore, Ava, a conservation project manager, suggested that hybrid cats may be better suited to the 'hybrid landscapes' (i.e., the anthropogenic environment of the Anthropocene) of which they are denizens than 'pure' wildcats: 'it looks a lot like hybrids are probably going to be much better at surviving out there than wildcats might be because we have hybrid landscapes, we have hybrid habitat out there, and hybrids survive in it and wildcats don't'[7]. This echoes Alfie's comments on habitat preference and the wildcat's anthropophobic nature, which puts it at a disadvantage in the Anthropocene.

While the lives of feline analogues as posited here might initially be valued for replacing the wildcat, they could potentially, over time, acquire their own (conservation) value if the number and genetic diversity of 'pure' wildcats decreases, if habitat destruction and fragmentation continues, and if conservationists 'embrace compound organisms and novel ecosystems' (Palmer et al., 2021, see also Schlaepfer et al., 2011). The point about embracing compound organisms and novel ecosystems is relevant to cats but also other species, and is particularly relevant to rewilding. Advocates and proponents of rewilding discuss its 'emergent properties' and suggest that it will require a 'radical tolerance' for other-than-human species and their agency. The major question with regard to rewilding in Britain is the extent to which people are prepared to tolerate emergent properties and the unruliness of other-than-human animals exerting their agency by hybridising (von Essen and Allen, 2016). Scholars conjure us to 'hang on' to that unruliness as offering 'a way forward in [… the] ecological revivification and repair' (Rutherford, 2018), which can occur in the margins (i.e., in hybridity and ferality). Indeed, scholars see ferality as 'generative' and 'full of potential' (Rutherford, 2018) – perhaps the same could be said of hybridity.

Qualifying wildness

Exploring the lives and deaths of wild, feral and hybrid cats in Britain reveals how wildness is categorised and challenged, produced and policed. Wildness can be seen to be profoundly blurred, and even contradictory, since the boundary between it and domesticness, and the boundaries between species, are porous and permeable. Wildness is, nonetheless, valorised, with conservationists valuing wildcats for their wildness and caring for them in elaborate and sophisticated ways. This care is, however, controlling, coercive and highly interventionist, including in imposing human control on wildcat feeding and breeding to the point that it begins to resemble domestication, and the 'wildness' of wildcats can be questioned.

Conversely, conservationists do not value feral cats, whose presence is seen to require 'management' since it not only threatens wildcat conservation but conservation itself, through the erosion of the wild/domestic boundary. This occurs most notably when wild and feral cats exert their unruly agency by hybridising. The resultant hybrids, literally and figuratively, resist easy classification, exposing the limitations of binary thinking in conservation.

Simultaneously viewed as monstrous and impure, and as highly adaptive members of our novel ecocultures, able to occupy an anthropogenic landscape to which they may be better suited than their wild counterparts, hybrids occupy a liminal ecological and ontological space. The intermingling of wild and domestic genes means that this hybridisation can be seen as another way in which the wildcat is being domesticated, or perhaps more accurately, engaging in 'socio-environmental co-evolution' (Stephanoff and Vigne). These hybrids are emblematic of dwelling, and attempting conservation, in the Anthropocene, where wildness is no longer a distinct concept but has many qualities and where we find ourselves in highly compromised and compromising positions which, nonetheless, have promise and potential.

Scholars describe conservation in such circumstances as 'gardening in the ruins' (Tsing, 2014), and while the term gardening suggests a level of intervention that is antithetical to rewilding theory, it reflects that rewilding practice is operating in real rather than ideal circumstances (*sensu* van Dooren, 2014) and therefore adopting a pragmatic approach. Thus, the care involved in rewilding can be seen as a practice 'that aims to nourish and sustain species and their living participants in far-from-ideal conditions, where the most desirable options simply are not available' (van Dooren, 2014, p. 116). Rewilding in the Anthropocene is thereby challenged to accommodate ambiguity, and humans are challenged to live with other-than-human animals whose unruly agency resists the categories and values we might hold or impose. We are challenged to consider what kinds of multispecies futures might be possible if we move beyond the valorisation of purity and wildness and embrace the messy complexities that wildness has come to contain.

Notes

1 It should be noted that some recent hybridisations which are occurring seemingly 'independently' of human intervention are causing consternation among conservationists since, while they may not be *directly* influenced by humans they are seen as the *indirect* result of human activity. Hybridisation between polar (*Ursus maritimus*) and grizzly (*Ursus arctos horribilis*) bears as they change their ranges due to anthropogenic climate change provides an excellent example of this, demonstrating the pervasive effects of the Anthropocene and lending weight to Head's (2015) claim that the environmental and social realms are no longer separable.

2 Feral and hybrid cats are considered overabundant both in absolute terms (their total number) and in notional terms: the value, or rather lack of value, assigned to these cats leads to the view that they should not exist at all, making their presence, in any number, an 'overabundance'.

3 A review of research into the post release survival of reintroduced carnivores suggests that the mortality of captive bred carnivores may be even higher: researchers found that the survival of captive bred carnivores post-release was 32% and that of wild-caught carnivores post-reintroduction was 50% (Jule, Leaver and Lea, 2008).

4 Feral cats are not valued due to their 'domesticness' while hybrids cats are not valued since their very existence is seen as 'unnatural' (*sensu* Brown and Sax, 2005) and as crossing the nature-culture boundary (Fredriksen, 2016), largely, though not wholly, due to what is perceived as human intervention in bringing about the hybridisation (Allendorf et al., 2001).

5 The vaccination element of these programmes is interesting in its care about and intervention in (and therefore arguably violence to) feral and hybrid cat immune systems. In the interests of simplicity, however, it is not discussed here.

6 This control involves confinement via trapping (albeit temporary), removal of reproductive agency via neutering (c.f. Srinivasan, 2013 who discusses this in relation to street dogs in India), removal of immunological agency via vaccination, and even, post-release, ongoing control by *othering* under the label of hybrid or feral (Johnston, 2021).

7 It should be noted that the wildcat's decline is multifactorial and includes persecution and hybridisation. Nonetheless, habitat loss is a significant factor and, as their critically endangered status attests, wildcats have struggled to thrive in anthropogenic landscapes.

References

Allendorf, F. W., Leary, R. F., Spruell, P. and Wenburg, J. K. (2001) 'The problems with hybrids: Setting conservation guidelines', *Trends in Ecology & Evolution*, 16(11), pp. 613–622.

Asdal, K., Druglitrø, T. and Hinchliffe, S. (2017) 'The 'more than human' condition: Sentient creatures and versions of biopolitics' in K. Asdal, T. Druglitrø and S. Hinchliffe (Eds) *Humans, Animals and Biopolitics*. London: Routledge.

Braverman, I. (2016) 'The regulatory life of threatened species lists' in I. Braverman (Ed.) *Animals, Biopolitics, Law: Lively Legalities*. London: Routledge, pp. 19–38.

Breitenmoser, U., Lanz, T., and Breitenmoser-Würsten, C. (2019) *Conservation of the Wildcat (Felis Silvestris) in Scotland: Review of the Conservation Status and Assessment of Conservation Activities*. Bern, Switzerland: IUCN SSC Cat Specialist Group.

Brown, J. H. and Sax, D. F. (2005) 'Biological invasions and scientific objectivity: Reply to Cassey', *Austral Ecology*, 30(4), pp. 481–483.

Fredriksen, A. (2016) 'Of wildcats and wild cats: Troubling species-based conservation in the Anthropocene', *Environment and Planning D: Society and Space*, 34(4), pp. 689–705.

Head, L. (2015) 'The anthropoceneans', *Geographical Research*, 53(3), pp. 313–320.

Hill, K., Szydlowski, M., Heaney, S. O. and Busby, D. (2022) 'Uncivilized behaviours: How humans wield 'feral' to assert power (and control) over other species', *Society and Animals, 31*(7), pp. 907–925.

Johnston, J. J. (2021) 'Imposing ferality: A technopolitical analysis of feral and free-roaming animal classification technologies', *Urban Geography*, 44(10), pp. 2163–2184.

Jule, K. R., Leaver, L. A. and Lea, S. E. G. (2008) 'The effects of captive experience on reintroduction survival in carnivores: A review and analysis', *Biological Conservation*, 141(2), pp. 355–363.

Katz, E. (1997) *Nature as Subject*. Oxford: Rowman and Littlefield.

Lorimer, J. and Driessen, C. (2016) 'From "nazi cows" to cosmopolitan "ecological engineers": Specifying rewilding through a history of Heck cattle', *Annals of the American Association of Geographers*, 106(3), pp. 631–652.

McDonald, J. L. and Skillings, E. (2021) 'Human influences shape the first spatially explicit national estimate of urban unowned cat abundance', *Scientific Reports*, 11, 20216.

Milton, K. (2000) 'Ducks out of water. Nature conservation as boundary maintenance' in J. Knight (Ed.) *Natural Enemies. People-Wildlife Conflicts in Anthropological Perspective*. London: Routledge, pp. 229–246.

Noddings, M. (1999) 'Two concepts of caring', *Philosophy of Education*, pp. 36–39.

Ottoni, C. et al. (2017) 'The palaeogenetics of cat dispersal in the ancient world', *Nature Ecology & Evolution*, 1(0139), pp. 1–7.

Palmer, A., Sommer, V. and Msindai, J. N. (2021) 'Hybrid apes in the Anthropocene: Burden or asset for conservation?', *People and Nature, 3*, pp. 573–586.

Palmer, A. and Thomas, V. (2023) 'Categorisation of cats: Managing boundary felids in Aotearoa New Zealand and Britain', *People and Nature, 5*, pp. 1539–1551.

Palmer, C. (2016) 'Climate change, ethics, and the wildness of wild animals' in B. Bovenkerk and J. Keulartz (Eds) *Animal Ethics in the Age of Humans: Blurring Boundaries in Human-animal Relationships*. Cham: Springer, pp. 131–150.

PDSA (2024) *How many pets are there in the UK?* Available at: https://www.pdsa.org.uk/what-we-do/pdsa-animal-wellbeing-report/uk-pet-populations-of-dogs-cats-and-rabbits (Accessed: 21 June 2025).

Rutherford, S. (2018) 'The Anthropocene's animal? Coywolves as feral cotravelers', *Environment and Planning E: Nature and Space*, 1(1-2), pp. 206–223.

Saving Wildcats (2023a) *Wildcats in Scotland*. Available at: https://www.savingwildcats.org.uk/about-saving-wildcats/wildcat-conservation-in-scotland/ (Accessed: 21 June 2025).

Saving Wildcats (2023b) *Released wildcats are being tracked with GPS radio collars. Available at*: https://www.savingwildcats.org.uk/news-events/2023/june/wildcats-tracked-with-gps-radio-collars/ (Accessed: 21 June 2025).

Scasta, J. D. (2019) 'Why are humans so emotional about feral horses? A spatiotemporal review of the psycho-ecological evidence with global implications', *Geoforum, 103*, pp. 171–175.

Schlaepfer, M. A., Sax, D. F. and Olden, J. D. (2011) 'The potential conservation value of non-native species', *Conservation Biology, 25*, pp. 428–437.

Srinivasan, K. (2013) 'The biopolitics of animal being and welfare: Dog control and care in the UK and India', *Transactions of the Institute of British Geographers, 38*, pp. 106–119.

Todesco, M. et al. (2016). 'Hybridization and extinction', *Evolutionary Applications, 9*, pp. 892–908.

Tsing, A. L. (2014) 'Blasted landscapes, and the gentle art of mushroom picking' in E. Kirksey (Ed), *The Multispecies Salon*. New York: Duke University Press, pp. 87–110.

van Dooren, T. (2014) *Flight Ways – Life and Loss at the Edge of Extinction*. New York: Columbia University Press.

von Essen, E. (2017) 'Whose discourse is it anyway? Understanding resistance through the rise of "Barstool Biology" in nature conservation', *Environmental Communication*, 11(4), pp. 470–489.

von Essen, E. and Allen, M. (2016) 'Wild, but not too-wild animals: Challenging Goldilocks standards in rewilding', *Between the Species*, 19(1), pp. 80–108.

Wagner, W., Kronberger, N., Nagata, M., Sen, R., Holtz, P. and Palacios, F. F. (2010) 'Essentialist theory of 'hybrids': From animal kinds to ethnic categories and race', *Asian Journal of Social Psychology, 13*(4), pp. 232–246.

Woods, M. (2005) 'Ecological restoration and the renewal of wildness and freedom', in Heyd, T. (ed.) *Recognizing the Autonomy of Nature: Theory and Practice*. New York: Columbia University Press, pp. 170–188.

Yamaguchi, N., Kitchener, A., Driscoll, C. and Nussberger, B. (2015) *Wildcat (Felis silvestris) The IUCN Red List of Threatened Species*. Available at: https://doi.org/10.2305/IUCN.UK.2015-2.RLTS.T60354712A50652361.en

Zeuner, F. E. (1963) *A History of Domesticated Animals*. London: Hutchinson.

8 Gods and monsters

The charismatic megafolk of rewilding

This chapter marks an important departure from the other chapters in part two of this book and considers the *human* animals involved in rewilding. The term *charismatic megafauna* is well known in conservation discourse to describe the large, other-than-human animals (usually, but not always, terrestrial mammals) that are compellingly attractive to humans. Charismatic *megafolk*[1] is extrapolated from this concept to describe the charismatic, often 'larger than life', humans involved in conservation. This chapter starts with a brief outline of charisma before going on to reflect on charisma, celebrity and charismatic megafolk in conservation and then charisma and charismatic megafolk in rewilding and reintroductions. It then examines the charismatic megafolk promoting rewilding in England and Britain, making overt their importance and the importance of charisma in rewilding. Three categories of charismatic megafolk are identified: *trendsetters, soapboxers* and *mavericks*. The category of mavericks is divided into two further categories: *guerrilla rewilders*, and *narcissists and Machiavellis*. These categories are discussed in turn before concluding with a brief reflection on charisma, its influence on rewilding, what this means and whether it matters.

A chapter such as this needs to be prefaced with several disclaimers. First, in discussing charismatic megafolk (of conservation and rewilding), this chapter focuses almost exclusively on white men. This is not to condone the lack of diversity in conservation and rewilding. The lack of diversity is, however, inescapable at the time of writing, and this chapter reflects the current, and indeed historical, situation, as has been identified by other authors: conservationists have generally been (and still are) white men and rewilding, for all its radical intent, has not avoided this. (Brockington, 2009; Huggan, 2013; Goldenberg; 2014; Abidin et al., 2020). Second, while the charismatic megafolk and their actions discussed here are all motivated by a desire to advance reintroductions and rewilding, the means used are not always legal and, whether legal or not, do not always achieve the intended effect. This chapter seeks to report on this in a symmetrical way (in which a sophisticated analytic perspective is attempted which, while not strictly neutral, makes that non-neutrality transparent and is not aligned with any of the perspectives under investigation (Sismondo, 2010)), it in no way seeks to

DOI: 10.4324/9781003456490-9

commend, condone or condemn any actor or action, nor to suggest that any particular individual or organisation has been involved in illegal or unethical activity. Third, and relatedly, in discussing sensitive topics, this chapter raises issues which could, potentially, result in reputational harm to the people concerned: for this reason, unless information is in the public domain, all names are omitted or changed and all quotes are left unattributed, this extends to people, locations and, where the mention could reveal the actors involved, species.

Charisma

Having *charisma* implies that someone possesses:

> a certain quality of individual personality by virtue of which he is considered extraordinary and treated as endowed with supernatural, superhuman, or at least specifically exceptional powers or qualities. These are such as are not accessible to the ordinary person, but are regarded as exemplary, and on the basis of them the individual concerned is treated as a "leader"
>
> (Weber, 1978, p. 241).

The possession of charisma, then, makes people and their ideas compelling to others, giving them a form of 'referent power' and making people inclined to treat them as leaders (Weber, 1978). The 'recognised authority' which charisma gives leaders is not necessarily *claimed by them* but is *bestowed upon them* by others (Brockington, 2009).

It should be noted, however, that charisma is not always virtuous. While it can be associated with altruism, self-confidence and sound judgement it can also be associated with narcissism, arrogance and poor judgment (Williams et al., 2021). Nor is charisma always wielded benignly. The attributes which make charismatic people compelling mean that they can, by extension, compel those who revere them to engage in extreme, sometimes destructive, violent behaviour even if it stems from ultimately good intentions, or indeed to engage in such behaviour themselves. Examples of individuals who display such behaviour are Achilles and Cuchulain, heroes who became monsters, or were at least capable of monstrous acts, during their feats of heroism, convinced of their righteousness (Weber, 1968). Thus, while charisma is most commonly associated with positive attributes, it should be noted that it can create not only gods but also monsters. 'Monsters' is deliberately inflammatory language, and while this chapter is concerned with the 'gods' that charisma creates, the dark side of charisma must also be discussed since it appears, if not in the form of monsters, then at least in the form of its unruly consequences. These unruly consequences are comprehensible in light of the power of charisma and the license it affords those who possess it.

Charisma and celebrity in conservation

Conservation has a history of enrolling and producing charismatic and/or celebrity individuals. This is, not least, because being 'charismatic, persuasive and generally larger than life' is an essential part of a conservationist's role (Brockington, 2009). Since conservation work often entails challenging normative views on growth and development and influencing policy, which in turn involves dealing with egotistic politicians, industry leaders and entrepreneurs, conservationists need the confidence and assurance which accompany and are gained from charisma (Brockington, 2009). While celebrity, as opposed to charisma, is less a focus of this chapter, it remains relevant. In part, this is because it can be difficult to separate charisma and celebrity in conservation since 'the line between a conservation celebrity and a charismatic conservationist is a thin one and easily crossed' (Brockington, 2009). It is also, however, because the relationship between conservation and celebrity has been well examined and offers a useful route to examine conservation and charisma, and indeed rewilding and charisma.

A crucial difference between celebrity and charisma is that, as mentioned above, charisma is a trait which is both possessed by individuals and *conferred on them by others*. Charisma, therefore, is generated through authentic encounters while celebrity is produced via contrived media exposure (Brockington, 2009). While celebrities may be charismatic, and charismatic individuals may be celebrities, the two are not necessarily correlated. There is, however, more likely to be a correlation between the charisma and celebrity *within conservation*. This is because, by its very nature, conservation generates (apparent) authenticity: in contrast with the egoistic pursuit of courting celebrity (which can create the impression of vanity and self-interest), championing conservation is ostensibly altruistic, lending credibility to those involved (Brockington, 2009). Moreover, conservationists often find themselves portrayed in ways that are at odds with the perfectly curated image which celebrity usually seeks to generate – this apparent lack of concern for cultivating a mediagenic profile lends conservationists an integrity and authenticity which in turn affords them charisma. Thus, conservationists are sometimes able to combine celebrity and charisma to significant effect. For example, in 2006, David Attenborough, a British naturalist and broadcaster, well known for presenting natural history and wildlife television programmes and for championing conservation, was considered the most trusted celebrity in Britain (Brockington, 2009). Attenborough's authority and appeal are attributed to his authenticity, which is generated by his behaviour: Attenborough being 'so prepossessed by his fascination with the subject at hand and unconcerned for his own dignity in front of the camera that he seemed to sweat integrity' (Cubitt quoted in Brockington, 2009).

Attenborough is cited in conservation literature as both charismatic *and* a celebrity (Brockington, 2009; Sullivan, 2011). More specifically, as a maker and presenter of wildlife and natural history programmes, Attenborough is described as a *celebrated conservationist*,[2] being a conservationist whose work in broadcasting makes them a celebrity and raises public awareness about, and concern for, the natural world, and the conservation issues they focus on. Steve Irwin, an Australian naturalist and broadcaster, is also cited as an example of a celebrated conservationist (Brockington, 2009; Sullivan, 2011). Like Attenborough, Irwin was highly charismatic and, again like Attenborough, this stemmed from, *inter alia*, his spontaneous presenting style, which was entirely counter to the carefully staged performances normally associated with celebrity, enhanced by Irwin's often 'worse for wear' appearance and seemingly genuine exclamations of 'crikey' (Brockington, 2009). In some ways, this makes Irwin the 'antithesis of Attenborough' (Brockington, 2009): while both men's (apparent) lack of concern for their own dignity affords them authenticity, they have very different styles in obtaining this authenticity. These different routes to authenticity and charisma become relevant later in discussing the charismatic megafolk of rewilding, as indeed, although to a lesser extent, do celebrated conservationists. Two other forms of celebrity in conservation are also important to mention here, as, again, they will be relevant to the discussion of celebrity in rewilding: *conservation celebrities* and *celebrity conservationists*. Conservation celebrities are those who have gained celebrity through their conservation work, while celebrity conservationists are those who have gained celebrity and then lend their support and, crucially, the influence which their celebrity affords them, to conservation causes (Brockington, 2009; Sullivan 2011).

To turn more specifically to charisma, conservation has long relied on charisma to further its cause, not only from conservationists but from the species it is seeking to protect. *Charismatic megafauna* is a recognised term for large animals which have characteristics that make them appealing and alluring to humans: these characteristics usually, although not always, include being terrestrial, mammalian and 'exotic'[3] (Thomas, 2022). Charismatic megafauna are often used to garner support for their own conservation, as an 'umbrella species'[4] to protect species and habitats more widely, and/or to promote broader conservation movements or organisations. *Charismatic megaflora* is a related, though less well recognised, concept which suggests that plants, or certainly trees, can play a similar role (Hall, James, & Baird, 2011). More pertinently, however, scholars have extended the concept of charismatic megafauna to include people, or rather celebrities, and the further blurring of celebrity and charisma is notable here (Boykoff, Goodman & Littler, 2010).

The distinction between human and other-than-human animals as charismatic megafauna is, however, an important one and is sufficiently important to warrant a new term – *charismatic megafolk*. Charismatic megafolk is used here to describe and discuss the charismatic people involved in conservation. As with charisma and celebrity, there may, or may not, be an overlap between charismatic megafolk and celebrity – some charismatic megafolk may have celebrity status while others may not. It should also be noted that the 'celebrity' of charismatic megafolk in conservation may be relatively small-scale – charismatic megafolk may be famous (or perhaps infamous) within conservation circles, but unknown to wider publics. This is, not least, because charisma, while being a characteristic, and while being associated with celebrity, can also be associated with wealth, power and status. Indeed, the influential role that 'charismatic conservationists' play in forming 'wealthy social networks' (Brockington, 2009) has been identified, emphasising the importance of wealthy (charismatic) individuals in supporting and, directing, conservation via 'well-connected and networked' 'conservation elites' who shape 'conservation discourses and practices' (Holmes, 2011). Charismatic conservationists and conservation elites can both be encompassed by the concept of charismatic megafolk and will be covered in the discussion in the following sections, making the case that celebrity, wealth and charismatic individuals are equally as, if not more, important to rewilding as they are to conservation more broadly.

Charisma and celebrity in rewilding

The shift from charisma and celebrity in *conservation*, to charisma and celebrity in *rewilding* turns on three points, two of which work in favour of the association and one of which complicates the relationship. Mediagenic landscapes and species are those which are most likely to be associated with celebrity, often the preservation of wilderness and charismatic species specifically (Brockington, 2008, 2009). Thus, rewilding, with its appeal for wilder landscapes and the reintroduction of keystone species and ecosystem engineers (which are often, although not always, charismatic megafauna), is ideally suited to endorsement from celebrity conservationists, attracting the attention of celebrated conservationists, and to creating conservation celebrities. Furthermore, celebrity conservation, whether by celebrity conservationists, celebrated conservationists or conservation celebrities, does not stop at simply conserving the world but 'actively creates and remakes it' Brockington (2009). Self-evidently, this, again, aligns very well with the principles of rewilding, which, rather than conserving the *status quo*, aim to restore ecosystems and species, imagining a possible future of mutual flourishing for people and nature.

The apparent mutualism between celebrity, charisma and rewilding is, however, complicated by the controversy surrounding rewilding. Compared to other causes that celebrities might champion, conservation (in its widest sense) is seen as uncontroversial, and conservationists are generally viewed as working towards the achievement of an intrinsic good (i.e., the conservation of nature, which publics are increasingly recognising is threatened and in need of protection) (Holmes, 2011). The controversy surrounding rewilding, on the other hand, is well established, with concerns over the exclusion of people, the reintroduction of species, changes to valued (and productive) landscapes, endangered species conservation and animal welfare (Thomas, 2021). What allows the celebrity-rewilding relationship to be maintained is that both celebrity conservation and rewilding have a tendency to deal in abstract narratives about 'remote' or 'wild' places, or to address species conservation and ecological restoration in very abstract terms (Brockington, 2008). Advocates of rewilding can deliberately seek to create or emphasise a sense of abstraction between rewilding discourse and specific rewilding sites since, as Charlie, a rewilding expert, explained, 'when you're talking globally it's not as scary as when you're talking about specific national instances'. The controversy associated with rewilding is also somewhat offset by its novel and radical approach and its ability to enthuse, inspire and excite, catching the popular imagination, thus restoring its compatibility with celebrity (Deary and Warren, 2018). Two celebrities can be highlighted here for their association with rewilding: Leonardo DiCaprio and Chris Packham.

Packham, an English naturalist, gained fame through his work as a television presenter of nature and wildlife programmes and thus is an example of a celebrated conservationist (Brockington, 2009; Sullivan, 2011). In June 2021, Packham added his name to an open letter calling on the British royal family to rewild their estates (Wildcard, 2021). Four months later, he led a demonstration in which he and a group of school strikers[5] delivered a petition (with over 100,000 signatures) to Buckingham Palace calling for the same thing (Bryant, 2021). Then, in December of the same year, Packham met representatives of the Crown Estate to discuss the proposal (Barkham, 2021). This activity made his name synonymous with rewilding in Britain, and this is being cemented by his more recent calls for the rewilding of parts of Glen Coe and the fact that he is presenting a new wildlife series from the Dundreggan Rewilding Centre (Davies and McKenzie, 2023). Interestingly, Packham's approach is, like Irwin's, being juxtaposed with Attenborough's. Packham's low-budget YouTube series is alluded to as radical, punk and controversial (all adjectives which have been used to describe rewilding (Thomas, 2021)) and contrasted with Attenborough's high-budget, and more conformist, BBC series.

DiCaprio, an American actor, gained fame through his acting career and has now lent it to conservation, making him an example of a celebrity conservationist (or celebrity rewilder). DiCaprio is associated with a range of environmental and conservation movements, but of interest here is his involvement, as a founding board member, in *Re:wild*, an organisation which seeks to protect and restore the wild. One of Re:wild's main activities is fundraising 'to address gaps in financing, kickstart careers and leverage impact to enable the conservation community to protect and restore the wild in the most powerful ways possible' (Re:wild, 2025a). Another is as a 'force multiplier that brings together indigenous peoples, local communities, influential leaders, nongovernmental organisations, governments, companies and the public to protect and rewild at the scale and speed we need' (Re:wild, 2025b). Thus, the work of DiCaprio and Re:wild, billed as 'the latest undertaking linked to DiCaprio's environmental activism' (Re:wild, 2025c) tallies almost exactly with the descriptions of charismatic/celebrity conservationists forming wealthy, well-connected, highly influential networks of elites to shape conservation (or in this case rewilding) discourse and practice (Brockington, 2009; Holmes, 2011).

It should be noted that wealth appears to be particularly important with respect to rewilding, perhaps not least because one of the earliest, and most enduring, tenets of rewilding is that it should occur at 'landscape scale' (Thomas, 2021). Achieving landscape-scale rewilding requires either the cooperation of several landowners or the inclination of owners of large tracts of land. Self-evidently, the latter usually necessitates that the landowner in question is wealthy and/or has the support of wealthy backers. Given the complexities of collaborating on land management, single-owner rewilding projects can be more straightforward, and decisions can be made more quickly. Well-known examples of rewilding projects instigated by wealthy landowners are *Rewilding Chile* and *Rewilding Argentina*, founded and supported by Kristine and Douglas Tompkins (Tompkins Conservation, 2025) and the Glenfeshie Estate (and other ventures) as part of Anders Holch Povlsen's *Wildland* vision (Wildland, 2021). Such people have been described as 'billionaire celebrity humanitarians who, although famous for their entrepreneurial success, deepen their cultural imprint through a range of environmental work run through their nonprofit organizations and foundations' (Abidin et al., 2020, p. 398). This is perhaps particularly true of Kristine Tompkins, who could now justly be considered a conversation (or rewilding) celebrity as a result of her work.

The wealth of people like the Tompkins and Povlsen allows them to have direct and indirect influence on rewilding. They can buy land on which to enact their rewilding vision directly, and, in doing so, they set land management trends, particularly through their networks with other wealthy, influential individuals, which influences other landowners and managers. Such influence is often seen as benevolent, as philanthropy seeking to support a worthwhile cause. There can, however, be concern that such philanthropists made their wealth through environmentally damaging activities (the Tomkins and Povlsen made considerable wealth in the clothing industry), and/or that they lead extravagant lifestyles which are environmentally damaging and that rather than being genuinely altruistic their philanthropy is an attempt to atone for such damage and assuage their attendant guilt. Furthermore, while the messaging from such philanthropic endeavours is that the rewilding they are advancing is for the good of the environment and for humanity, there is often the suspicion that it is actually for their own benefit – either that they are guarding the environment for themselves or the glory of their legacy (Huggan, 2013).

The charismatic megafolk of rewilding – gods and monsters

Having discussed charisma and celebrity in relation to conservation and rewilding, this chapter now goes a step further to discuss the charismatic megafolk of rewilding in Britain. The crux of the concept of charismatic megafolk is that, like charismatic megafauna (and indeed, though to a slightly to a lesser extent, charismatic megaflora), they act as ambassadors in promoting rewilding in both conservation and public discourse. During the course of the research for this book, one participant asked: 'is it organisations who make the difference in conservation or is it individuals?' The participant was referring to the pioneering conservationists who have advanced conservation agendas nationally and globally, sometimes, but not always, while, or despite, working within a conservation organisation. Roy Dennis is an excellent example of one such pioneering conservationist. Dennis is an English ornithologist who has spent his career in the conservation of birds, including, and especially, raptors. During his career, Dennis has been involved with and employed by the Royal Society for the Protection of Birds (RSPB). Despite his association with the RSPB, however, Dennis has, in the most respectful terms, bemoaned the conservative nature of the organisation and worked within and without it to progress what the RSPB would have considered radical conservation methods, namely reintroductions (Dennis, 2021). It is reasonable to conclude that it is Dennis himself (as an individual) as much as the RSPB (as an organisation) who can take considerable credit for advancing the conservation agenda in Britain, particularly in relation to reintroductions, and in changing attitudes and approaches at the RSPB.

The role of (charismatic) individuals as opposed to organisations is perhaps even more important in the case of 'controversial' (or at least unconventional, somewhat radical) approaches such as rewilding and reintroductions. This is because individuals are less constrained by funding, members and protocols than organisations might be. It is also the case that charismatic individuals may be more prevalent in rewilding as compared with conservation since, as another participant observed: 'personalities are a big thing in this game, in conservation, but I think in rewilding especially, particularly as those doing rewilding now tend to be the bigger, brash ones anyway 'cause they're the ones bold enough to step forward and do something different'.[6] The identification of a predominance of 'big', 'bold' and 'brash' personalities in rewilding, and a recognition of the importance of the personalities of the individuals involved, feeds into the concept of charisma and charismatic megafolk. What is of interest here is that, as with other conservationists who have to challenge accepted norms and conventions, charismatic megafolk can leverage the traits and authority associated with charisma to shape, direct and advance rewilding agendas. This is particularly important, and apposite, with rewilding because significant force of will and personality is needed in order to persuade people to buy into rewilding's radical agenda and the significant re-visioning of the world entailed (Huggan, 2013).

Another role of charismatic megafolk (like celebrity conservationists) is in acting as a link between civilisation denuded of wildlife, and nature and the promise of regeneration (Brockington, 2009). Many people experience a disconnection from nature and crave the connection with nature that charismatic megafolk (and celebrity conservationists) enjoy. We can, however, experience nature vicariously through charismatic megafolk or by connecting with them and their rewilding projects (as will be discussed in the case of 'trendsetters' and associated ecotourism) (Brockington, 2008). Trendsetters are one of three species of charismatic megafolk identified here, the other two being 'soapboxers' and 'mavericks'. Mavericks can be divided into two sub-species: 'guerrilla rewilders' and 'narcissists and Machiavellis'. It should be noted that, by their very nature, the charismatic megafolk of rewilding advance, or at least seek to advance, the rewilding agenda. Of the categories identified, however, narcissists and Machiavellis and, to a lesser extent, guerrilla rewilders can, in fact, as will be discussed, hinder rewilding progress. It should also be noted that the charismatic megafolk described and discussed here were identified in relation to rewilding in Britain; it is, however, entirely plausible that similar charismatic megafolk can be found in relation to rewilding (and wider conservation) in other parts of the world.

Rewilding trendsetters

The first species of rewilding charismatic megafolk is the trendsetter.[7] Trendsetters are perhaps most similar to celebrity conservationists, and the archetypes here are Isabella Tree and Charles Burrell, who have been referred to as the king and queen of rewilding (Weston, 2022). Tree and Burrell derive their charisma, at least in part, from their wealth and social status; to misquote Stoppard, 'with rewilding as with divorce, there is a certain cachet in citing a member of the aristocracy' (1993, p.56): Burrell's status as the 10th Burrell Baronet gives him definite cachet in rewilding circles which reinforces, and is reinforced by, his wealth.[8] As discussed above, wealth, as a means of accessing large tracts of land, is particularly significant in relation to rewilding, as are the elite networks it fosters.[9] This is clearly illustrated by 'The Times' newspaper dubbing one such network 'the really wild billionaires club', running an article titled: 'Meet the Knepp group: the billionaires dedicated to "rewilding" the environment', accompanied by a picture of the group at Burrell and Tree's Knepp Estate (which they are rewilding) (Wade, 2018). The group exemplify conservation elites: 'actors who have global-scale (or at least continental) goals, and who can move influence, money, discourses and other resources of power around the world with relative ease' (Holmes, 2011, p. 2). Trendsetters are the charismatic megafolk who perform such a role for rewilding: they have grand rewilding ambitions and the means to carry them out. The link between rewilding trendsetting and wealth (and also status) is highlighted by a quote from a research participant:

> all the rewilding projects which exist, exist because one rich landowner has chosen to do them. Although you hear a lot of talk about 'wouldn't it be nice if we can get lots of small farmers to engage in rewilding, commoners[10] and the like' you won't because that requires all the commoners to agree on a certain approach. Whereas the rewilding scene at

> the moment requires one mind with control over a big area like Charlie Burrell, it's someone who's rich enough to go and buy a lot of land and have the vision and drive to go and make a significant change. Part of what Eve's[11] trying to do is create that commoner cooperation to some extent, but it's happening a lot more slowly than the toffs end of the market.

Because of the confidence associated with their charisma and perhaps, more pragmatically, because of their wealth, trendsetters can be, and have been, brave, embracing and adopting rewilding, and becoming its pioneers. In doing so, trendsetters make rewilding fashionable not only among their elite networks but to wider publics, although the ability to respond to and follow the trend may be very different. It is notable, however, that trendsetters can extend rewilding discourse beyond the usual conservation circle: an account, by Tree (2020), of rewilding the Knepp Estate, was published fashion magazine 'Vogue', and Tatler, another fashion magazine, included a rewilded estate on its list of 'modern status symbols', specifically citing Burrell and Tree as having started the trend (Sampson, 2022) (something reinforced by other landowners rewilding their estates and citing Tree and Burrell as influential in their decision making (Youens, 2022)). Such articles not only highlight Tree and Burrell's role as trendsetters but also highlight the role and influence of trendsetters more generally. In becoming pioneers of rewilding, and because of the allure associated with their charisma (together with wealth or status, which they may also possess), trendsetters not only demonstrate that it might be possible to manage land differently but make doing so appear desirable. Trendsetters thus create a virtuous circle for rewilding: by making rewilding fashionable, trendsetters lead others to adopt it, further driving the trend and increasing rewilding's popularity and momentum. Thus, trendsetters have considerable influence on rewilding both directly and indirectly: they influence rewilding directly through rewilding their own estates, but have perhaps even greater impact on rewilding indirectly, through the influence they have on other landowners in making a rewilded estate a status symbol.

Despite the ostensible altruism of trendsetters (which can, however, as discussed above, be questioned as to whether they are rewilding for the sake of the environment or for their own benefit (Huggan, 2013)), there is, nonetheless, a 'powerful economic logic' behind trendsetters' motives (Brockington, 2008). Being a rewilding trendsetter is a business requiring self-promotion: appearing in public and on news, popular and social media, giving talks, promoting and selling commodities (such as books, e.g., Tree's *Wilding* (2018)), and, perhaps most importantly, promoting and selling *experiences* (Brockington, 2008). Celebrity conservationists have been identified as offering 'closeness to nature' (Brockington, 2008), and rewilding trendsetters are no different: they offer us not only vicarious intimacy with nature but, through their ecotourism businesses, they afford us access, albeit temporary, to the kind of relationship with nature that they enjoy. Importantly, however, an additional part of the appeal is not only the access to nature but the potential access, even if unrealised, to the charismatic megafolk themselves: in going on a 'wildlife safari' on the Knepp Estate, ecotourists not only experience rewilding but also proximity to rewilding 'royalty'. And Knepp is not alone in hosting ecotourism associated with rewilding; ecotourism and rewilding of country estates go hand in hand since, not only are country estates particularly suited to rewilding, they are also peculiarly suited to ecotourism as attractive and desirable destinations.

Rewilding soapboxers

The second species of rewilding charismatic megafolk is the soapboxer.[12] Soapboxers, like trendsetters, are pioneers of rewilding, but rather than practicing rewilding (perhaps because they lack the means or the expertise), they use their talents (including their charisma) to act as vocal proponents for it. The archetype here is George Monbiot. who is famous, if not infamous, for sparking debate about rewilding in Britain. Monbiot's charisma and activism have been noted elsewhere in the literature on charisma in conservation. Monbiot has been described as 'radical' and included on a list of 'charismatic conservationists and environmentalists' (Brockington, 2009). Other scholars have identified 'activist intellectual' as one of the tropes of (charismatic) celebrity environmentalism and given Monbiot as an example (Abidin et al., 2020). While it is noted that activist intellectuals may come from many walks of life, specific mention is made of authors and journalists, such as Monbiot, 'who have gained celebrity and *notoriety* as outspoken environmental activists, campaigners, and speakers' (Abidin et al. 2020, p. 399 emphasis added). Activist intellectuals are strident critics of environmental issues who stage 'cerebral interventions' including by producing 'articulate reasoned and impassioned book-length arguments against the ills they are fighting' (Abidin et al., 2020, p. 393). Monbiot is thus a case in point, his book on rewilding, *Feral* (2013), having propelled rewilding into public consciousness in Britain, and Monbiot continues to contribute to the discourse through his comments in the news and social media, not least his column in 'The Guardian' newspaper (Thomas, 2021). Like trendsetters, soapboxers are important in raising the profile of rewilding as a radical, and controversial, concept: indeed, soapboxers sometimes deliberately provoke controversy in order, in their minds, to stimulate the rewilding debate. While this approach undeniably promotes discussion about rewilding, whether that discussion is constructive is less certain, having been criticised for 'polarising and antagonising stakeholders' (Sandom and Wynne-Jones, 2019, p. 222).

This polarisation and antagonisation provoked by soapboxers is partly due to the fact that they can be deliberately polemical, calling for radical action and adopting uncompromising positions (Brockington, 2009). Thus, while earlier in this chapter it was discussed that (with the exception of mavericks) charismatic megafolk advance the rewilding agenda, in deliberately provoking controversy, soapboxers may also retard rewilding progress, at least temporarily and at least in some quarters. For example, the controversy that Monbiot and other soapboxes provoked has led some rewilding practitioners in Britain to avoid using the term rewilding (Thomas, 2021). The counterargument would be that, while soapboxers may alienate some stakeholders, this is outweighed by the support and enthusiasm for rewilding which they generate in others. Soapboxers are, essentially, popularisers, using their charisma to appeal to and engage with wider publics and to create compelling narratives surrounding rewilding. These narratives, and soapboxers' appeal to emotions rather than reason, can be extremely effective in communicating rewilding.

The narrative they tell is thus central to soapboxers: given that they are involved in the discourse rather than the practise of rewilding and given that they are trying to enrol others in this discourse, their narrative and its authenticity, is crucial. One of the criticisms which could be levelled at soapboxers and other actors involved in promoting rewilding (e.g., armchair rewilders (see Thomas, 2022)) is that they do not have a legitimate right to comment on rewilding since they are distant from its activities and impacts (Thomas, 2022). Thus, the challenge for soapboxers is to legitimise their narrative and their right to tell it. To meet this challenge, soapboxers build 'crisis narratives' as a means by which they 'claim rights to stewardship over land and resources they do not own. By generating and appealing to crisis narratives, they assert rights as 'stakeholders' in the land and resources they say are under crisis' (Roe, 1995, p. 1066). In creating and controlling these narratives, soapboxers are drawing on a long history of narrative creation (or 'truth claims' (Abidin et al., 2020)) by charismatic/celebrity environmentalists. Early spokespeople for environmentalism were 'gurus' or 'prophets', such as Aldo Leopold and John Muir (Abidin et al., 2020), both of whom still feature in rewilding narratives and, in the case of John Muir, could be said to be influencing rewilding through the work of the John Muir Trust (2025).

Crucially, in creating these (crisis) narratives, soapboxers and others in the media can control how conservation issues are portrayed (Holmes, 2010). As discussed above regarding charisma and celebrity in rewilding, charismatic megafolk, including soapboxers, have a tendency to (over) simplify the social contexts within which rewilding operates, portraying 'depoliticized versions of nature and animals' (Brockington, 2008, p. 557). Thus, soapboxers can establish a narrative of rewilding being intrinsically good and worthy of support. They can also dictate the terms by which rewilding success is measured and, in doing so, they set the terms of the debate, allowing them to further control the discourse in order to advance rewilding and reinforce their status as spokespeople for it (Holmes, 2010). Such narratives produce and maintain 'conservationism' – an 'ideology' and a 'network of discursive practices' through which the practical measures of conservation are determined and endorsed (Huggan, 2013). Thus, through these narratives, soapboxers (attempt to) set the discursive practice through which the practical measures of rewilding are determined and endorsed.

Maverick rewilders

The third species of charismatic megafolk of rewilding is the maverick. Mavericks can be divided into two subspecies: guerrilla rewilders and narcissists, and Machiavellis. Collectively, mavericks are the more unconventional charismatic megafolk of rewilding. On one level, they are similar to some celebrity conservationists/rewilders, Steve Irwin or Chris Packham, for example, in that their personas are quite different from the more conventional, establishment figures of conservation or rewilding. Their spontaneous style, down-to-Earth approach and 'can do' (to the point of 'just get on with it') attitude sets mavericks apart from other conservationists/rewilders (including some other charismatic megafolk) who take a more reserved, cautious and conservative approach. Indeed, mavericks and maverick approaches are something of a reaction to 'the bureaucratization of conservation, mired in red tape, procedures, policy' (Brockington, 2009). Mavericks attempt to overcome or circumvent this bureaucratisation, which they see as an unnecessary barrier to rewilding progress. Over time, however, mavericks may themselves become mired in red tape and 'domesticated into the routine of conservation bureaucracies' (Brockington, 2009). Thus, mavericks are constantly replaced as new mavericks emerge to continue to challenge the bureaucracies which erstwhile mavericks have been incorporated into.

To a certain extent, mavericks fall into the seemingly oxymoronic category within charismatic or celebrity conservation of 'ordinary people' (Abidin et al., 2020). In this context, ordinary people are conservationists who are not, at least initially, members of 'elite' networks (membership of which is usually through celebrity, wealth or status: social or professional). 'Ordinary' people can, however, gain attention through their conservation work, becoming 'folk heroes'[13] defending the environment on our behalf – a classic example in environmentalism might be Swedish environmental activist Greta Thunberg (Brockington, 2009). We champion such heroes, partly because we see them acting on our behalf and partly because, in some ways, we like to imagine that we too are capable of performing such deeds (and achieving the same connection with nature) because, on some levels, they are just like us. Despite being 'ordinary', mavericks are, undoubtedly, charismatic and may even be tapping into and/or benefitting from the 'Steve Irwin effect' (Sullivan, 2011). This is a phenomenon attributed to Irwin's impact on media and public perceptions of conservation and 'the proliferation of a particular performance of nature, attracting attention through drama and sensation' (Sullivan, 2011). Mavericks certainly subscribe to the idea of attracting attention to rewilding through drama, sensation and through their performances and activities as charismatic megafolk.

Guerrilla rewilders

Guerrilla rewilders are the mavericks who perform guerrilla rewilding. Guerrilla rewilding is 'covert', 'renegade' or 'rogue' rewilding, commonly in the form of species reintroductions, which is undertaken illicitly (and, on occasions, illegally) (Thomas, 2022). The major motivation of guerrilla rewilders is frustration at the amount of time associated with the bureaucratic process of gaining official authorisation for rewilding projects (Thomas, 2022). Given the illicit (sometimes illegal) nature of guerrilla rewilding, guerrilla rewilders, understandably, conceal their identities – naming a guerrilla rewilder as an exemplar of this category is therefore neither practicable nor desirable. In the absence of a genuine example, Derek Gow, a Scottish farmer turned rewilder, will be used as a stand-in to discuss this category: this is in no way to suggest that Gow is a guerrilla rewilder nor to suggest that he has been involved with any guerrilla rewilding activities. Gow is, however, a maverick and, as one of rewilding in Britain's most vociferous advocates, he captures media and public attention due, *inter alia*, to his enthusiasm and charisma. Relying on Gow to stand in for anonymous guerrilla rewilders is not without precedent. In 2020, 'The Guardian' newspaper ran an article titled 'It's going to be our way now: the guerrilla rewilder shaking up British farming'[14] focussing on Gow's own (sanctioned) rewilding project and accompanying the article with a large image of Gow (Weston, 2020). At no point does the article suggest that Gow is involved in illicit or illegal rewilding activity, merely that he is a something of a renegade who takes an unconventional approach to rewilding, describing him as 'blunt' and 'determined', a 'visionary' to ecologists[15] (c.f. the idea of folk-heroes discussed above) and 'a pain in the arse' to government officials, and with a preternatural penchant for swearing (c.f. Irwin's everyday exclamations) (Weston, 2020). Another newspaper article, this time in 'The Sunday Times', describes Gow as a 'renegade rewilder' with little time for rules and a habit of 'bulldozing' his way through opposition.

Like folk-heroes, some of whom are 'outlaws', guerrilla rewilders take direct action, often transgressing laws and claiming that it is legitimate for them to do so since they are acting for altruistic reasons (a key determinant of charisma). Society grants folk-heroes licence to transgress laws on our behalf (indeed, the fact that we do so is what elevates them to folk-hero status), and we may grant the same latitude to guerrilla rewilders. Indeed, to some Gow, and by extension guerrilla rewilders, are visionaries (Weston, 2020). To others, however, they are dangerous radicals whose actions can jeopardise sanctioned rewilding projects and compromise public trust in and support for rewilding generally. This is because, while guerrilla rewilders decry conservation bureaucracy as a hinderance, to others it has an important, moderating influence (e.g., in adhering to the precautionary principle (Cooney, 2004)) without which charismatic megafolk and other conservationists can be drawn into extreme action (Brockington, 2009; also c.f. Webber's comments on how 'charisma may lead to excesses of derangement and deviance' (1968, p. xx) and to heroes such as Achillies and Cuchulain becoming monsters). While in the case of other charismatic conservationists, such extreme action is rare (Brockington, 2009), in the case of guerrilla rewilders, it is their *modus operandi*, and such direct action can have significant (intended or unintended) consequences on socioecological systems.

The transgressive behaviour that guerrilla rewilders exhibit arises from the fact that they are an extreme example of the charismatic conservationists for, which the bureaucracy of conservation simultaneously fuels demand and supply. Viewed positively, when empowered and emboldened by the limiting effect of bureaucracy on rewilding progress, guerrilla rewilders have the ability to 'bypass and evade' bureaucracies, and their rebellious attitudes and actions, and their 'immediacy' gives them a public appeal (Brockington, 2009). Indeed, even some conservationists recognise the beneficial effect that guerrilla rewilding can have in advancing the discourse and practice of rewilding since their actions can provide 'proof of concept' that rewilding in a particular area or with a particular species might work. For example, Gow (2023) has stated on social media that 'without unlicensed beaver release nothing would have been accomplished with this species in Britain at all', a claim which is supported by literature on the subject (Crowley, Hinchliffe and McDonald, 2017). Thus, even though the actions of guerrilla rewilders can be extreme there is an argument that their conduct does not necessarily reduce the significance of their achievements: if the efforts of guerrilla rewilders are successful (either directly, in and of themselves, or indirectly by advancing the rewilding agenda), the ends may justify the means of achieving them (Brockington, 2009). Indeed, there is a larger argument that charismatic megafolk such as guerrilla rewilders, and also soapboxers, have an important role to play in rewilding discourse and practice. Their radical and uncompromising attitudes can shift mainstream attitudes by expanding the 'Overton window' (Mackinac Centre for Public Policy (2019)) regarding rewilding, allowing more formal, political debate to follow and capitalise on the radical and controversial polemic, and actions, of soapboxers and guerrilla rewilders respectively.

Narcissistic and Machiavellian rewilders

The last of rewilding's charismatic megafolk is interesting in that they are different from the other charismatic megafolk discussed. As with guerrilla rewilders, an exemplar of a narcissist/Machiavelli will not be named, given the potential reputational harm that might ensue from doing so. While undeniably charismatic, narcissists and Machiavellis exemplify *dark* charisma – charisma which is influenced by the 'dark triad' of personality traits (narcissism, Machiavellianism and psychopathy) and thus manifests negatively rather than positively (Paulhus and Williams, 2002; Williams et al., 2021).[14] Charisma and dark charisma are two sides of the same coin: charismatic individuals might be described as confident and altruistic, while those with dark charisma could be described as arrogant and egoistic. Moreover, while charisma can support sound judgement and responsible behaviour, dark charisma can promote poor judgement and irresponsible behaviour. Nonetheless, darkly charismatic individuals are compelling in the same way that charismatic individuals are. As a result, they are equally able to appeal to, and even enthral, their audiences.

That being said, the selfishness and vanity associated with the dark triad personality traits can influence the judgment of those with dark charisma, leading them towards behaviours which are unethical and destructive. Thus, while like other rewilding charismatic megafolk, narcissists and Machiavellis may desire the advancement of rewilding, they can, unintentionally, retard or even sabotage rewilding progress rather than advancing it. It should be stressed that such sabotage may not be intentional and may not even be through actions which damage rewilding per se, but through disrupting rewilding discourse. Since narcissists and Machiavellis are as likely to alienate conservationists and rewilding stakeholders as they are wider publics, they can create dissent within the rewilding debate which can foster a wider public or political perception that there is uncertainty as to the benefits of rewilding, making policy makers and publics less likely to support it: two quotes from the research illustrate this clearly. The first illustrates the direct impact narcissists and Machiavellis can have on individual rewilding projects and those local to them: 'absolutely the local reception to rewilding is largely negative, partially cause it's something new, but I think also more so Adam[15] and his personality'. The second illustrates the much wider, indirect impacts which narcissists and Machiavellis can have on rewilding more generally:

> I think he actually made a lot of potential reintroductions nearly impossible with how he behaved and I think the entire conservation community is slightly pissed at him for what he did 'cause I think he fucked it for everyone else and he was so charismatic in the media: 'it's going to be amazing, it's going to be this', so you have loads of people in the south of the country, 'oh my God this is amazing', living in their apartments in London being, 'this is great, we're going to have wolves,[16] we can go on holiday', and the reality for the farmers that are actually there it was a terrifying prospect.

This quote is particularly interesting in that it highlights how narcissists and Machiavellis can alienate other conservationists and rewilding stakeholders (e.g., farmers) while simultaneously appealing to wider publics, with the participant specifically identifying the charisma of the individual in question. Perhaps even more importantly, this quote also suggests that this individual has hindered rewilding progress, albeit unintentionally.

Charisma's influence in rewilding

The charismatic megafolk of rewilding identified and described here (trendsetters, soapboxers, guerrilla rewilders, and narcissists and Machiavellis) represent a new understanding of the role of charisma in conservation, specifically in relation to rewilding. Charismatic megafolk are undeniably influential in setting and advancing the rewilding agenda: the important question which remains is to what extent it matters that a relatively small number of highly influential individuals are shaping the trajectory of rewilding. In rewilding their estates and in influencing other landowners to follow suit, trendsetters are directly and indirectly affecting how land is used. What is notable is that, as private owners of large areas of land, trendsetters and their elite networks can make land management decisions with little or no consultation with others who might reasonably be considered stakeholders in the land, for example, local communities or recreational visitors. Thus, a small number of people have a disproportionate influence on how a large proportion of land is managed, based predominantly on their individual agendas (Holmes, 2010). Similarly, in setting the discourse regarding rewilding, soapboxers can have a direct influence on how rewilding is portrayed in news, popular and social media, followed by a significant, indirect influence on how rewilding is perceived by those who consume that media. In taking direct action, guerrilla rewilders have a direct impact on how rewilding occurs and, given the nature of their activities, they too influence rewilding without consulting with other stakeholders. Narcissists and Machiavellis also have significant effects on rewilding, albeit not necessarily in advancing it but rather in stimulating controversy surrounding it. Collectively, then, charismatic megafolk have a disproportionate influence on rewilding discourse and practice compared with others in society.

'Elite' groups having major influence on conservation issues is not a new phenomenon (Holmes, 2010). It may, however, be becoming increasingly problematic in a world calling for more participatory decision making, particularly in relation to major environmental issues which have the potential to affect us all, thus making us all stakeholders in the outcomes of such decisions. Rewilding may therefore be vulnerable as a result of being so heavily reliant on decision-making by a small number of charismatic megafolk. Furthermore, if rewilding's future is heavily invested in charismatic megafolk, it is exposing itself to significant threats. Trendsetters and their followers may move on to a new fashion, abandoning rewilding. Soapboxers may adopt a new issue, leading their audiences with them. Guerrilla rewilders may take up a new cause to champion, leaving the rewilding they have started to run its own course. Indeed, guerrilla rewilders may have to cease their activities for other reasons if, for example, they attract too much attention or are identified and, potentially, charged with illegal activities. On a more sombre note, if particular rewilding projects are essentially dependent on individuals (as many of them are), then those projects could be under serious threat if anything happens to the individuals involved. Thus, while charismatic megafolk have a significant influence on rewilding, and while this influence can be beneficial, its extent may, for a number of reasons, be problematic.

Notes

1 The term charismatic megafolk is used courtesy of an anonymous research participant. They used the term in conversation following a research interview and it is used here with their knowledge and permission.
2 This chapter draws extensively on Brockington (2009) and Sullivan (2011), which is a reading of Brockington. The terms *celebrated conservationist, conservation celebrity* and *celebrity conservationists* are all Brockington's or Sullivans's and this chapter seeks to use them consistently according to the definitions given which are, again, derived from Brockington (2009) and Sullivan (2011).
3 It is notable that while terrestrial and mammalian are objective assessments, exoticism is very much a subjective evaluation. It should also be that noted that 'large' can be a relative term and that an animal may be considered charismatic megafauna if it is 'large' in relation to other representatives of the same category. For example, if charisma in recognised in birds as well as mammals, the red kite may be considered charismatic because of its size relative to other birds.
4 Umbrella species are, usually, charismatic species or megafauna which are protected for their own sake but also on the assumption that preserving them and their habitat offers 'a wide umbrella of land protection under which many species that are more abundant but smaller and less charismatic find safety and resources' (Soule and Noss, 1998, p. 21).
5 School strikers are children around the world who boycott school lessons on Fridays and instead use the time to call for action on climate change. The School Strike for Climate movement was initiated by environmental activist Greta Thunberg, arguably another conservation celebrity, although environmental celebrity may be more accurate.
6 This participant did not intend any offence in using the term 'brash'. While brash can denote arrogance and aggressiveness when carried to excess, in this context it was used with the positive connotations of confidence, assertiveness and boldness associated with charisma.
7 To some extent there is an overlap between trendsetters and the 'pioneer farmers' discussed elsewhere as proponents of rewilding, although the overlap is not exact (Thomas, 2022).
8 Another example of aristocracy associated with rewilding is Randal Plunkett, 21st Baron of Dunsany who is rewilding his estate in County Meath, Ireland.
9 The presence of headlines such as that in 'The Guardian' in 2022 of 'Rewilding 'not just for toffs' as one in five councils in Great Britain get onboard' (Greenfield, Yeo and Grant, 2022) illustrates how closely associated rewilding in Britain is, or at least has been, with wealth and aristocracy.
10 This participant is using the term 'commoner' to refer to those with access to common land, rather than in the sense of commoners as opposed to aristocracy. Given the context here, however, the latter reading allows for an interesting consideration of rewilding by 'commoners' as opposed to rewilding by 'toffs'.
11 Not her real name.
12 Again, there is some overlap between soapboxers and other proponents of rewilding discussed elsewhere (namely armchair rewilders and policy entrepreneurs) although again the overlap is not exact (Thomas, 2022).
13 Folk heroes are 'ordinary' people who, through their deeds (often carried out on behalf of other 'ordinary' people) gain heroic status, the classic example in English folklore being Robin Hood (Klapp, 1949).
14 Williams et al. give a very helpful explanation of the dark triad personality traits: '(1) narcissism: an inflated self-view and entitlement; (2) Machiavellianism: lacking concern for others, manipulative, and a moral outlook that puts the ends ahead of the means; and (3) psychopathy: having a disregard for people and social norms and lacking guilt' (2021).

15 Not his real name.
16 The species in question has been changed here in order to protect the subject of the debate and therefore the identity of the individuals involved.

References

Abidin, C., Brockington, D., Goodman, M. K., Mostafanezhad, M. and Richey, L. A. (2020) 'The tropes of celebrity environmentalism', *Annual Review of Environment and Resources*, 45, pp. 385–411.

Barkham, P. (2021) 'Chris Packham meets Crown Estate to promote rewilding royal land', *The Guardian*, 16 December. Available at: https://www.theguardian.com/uk-news/2021/dec/16/chris-packham-meets-crown-estate-to-promote-rewilding-royal-land

Boykoff, M., Goodman, M. and Littler, J. (2010) 'Charismatic megafauna: The growing power of celebrities and pop culture in climate change campaigns', Environment, Politics and Development working paper 28, London: King's College Department of Geography.

Brockington, D. (2008) 'Powerful environmentalisms: Conservation, celebrity and capitalism', *Media, Culture & Society*, 30(4), pp. 551–568.

Brockington, D. (2009) *Celebrity and the Environment: Fame, Wealth and Power in Conservation*. London: Zed Books.

Bryant, M. (2021) 'Chris Packham asks royal family to commit to rewilding estates', *The Guardian*, 9 October, Available at: https://www.theguardian.com/environment/2021/oct/09/chris-packham-asks-royal-family-to-commit-to-rewilding-estates

Cooney, R. (2004) *The Precautionary Principle in Biodiversity Conservation and Natural Resource Management: An Issues Paper for Policy-makers, Researchers and Practitioners*. Gland: IUCN.

Crowley, S. L., Hinchliffe, S. and McDonald, R. A. (2017) 'Nonhuman citizens on trial: The ecological politics of a beaver reintroduction', *Environment and Planning A*, 49(8), pp. 1846–1866.

Davies, E. and McKenzie, S. (2023) 'Chris Packham: Glen Coe is broken and needs fixing', *BBC Scotland*, 25 October. Available at: https://www.bbc.com/news/uk-scotland-highlands-islands-67204307

Deary, H. and Warren, C. R. (2018) 'Trajectories of rewilding: A taxonomy of wildland management', *Journal of Environmental Planning and Management*, 62(3), pp. 466–491.

Dennis, R. (2021) *Restoring the Wild: Sixty Years of Rewilding Our Skies, Woods and Waterways*. London: Harper Collins.

Goldenberg, S. (2014) 'Why are so many white men trying to save the planet without the rest of us?' *The Guardian*, 8 May. Available at: https://www.theguardian.com/commentisfree/2014/may/08/white-men-environmental-movement-leadership

Gow, D. [@gow_derek]. (2023, June 13). Slow plodding practice meets head on with the rising desperation for action. Without unlicensed beaver release nothing would have been accomplished. [X post]. X. https://x.com/gow_derek/status/1668572145813794817

Greenfield, P., Yeo, S., and Grant, C. (2022) 'Rewilding 'not just for toffs' as one in five councils in Great Britain get onboard', *The Guardian*, 20 June. Available at: https://www.theguardian.com/environment/2022/jun/20/rewilding-not-just-for-toffs-as-one-in-five-councils-in-great-britain-gets-onboard-aoe

Hall, C. M., James, M. and Baird, T. (2011) 'Forests and trees as charismatic mega-flora: Implications for heritage tourism and conservation', *Journal of Heritage Tourism*, 6(4), pp. 309–323.

Holmes, G. (2010) 'The rich, the powerful and the endangered: Conservation elites, networks and the Dominican Republic', *Antipode*, 42(3), pp. 624–646.

Holmes, G. (2011) 'Conservation's friends in high places: Neoliberalism, networks, and the transnational conservation elite', *Global Environmental Politics*, 11(4), pp. 1–21.
Huggan, G. (2013) *Nature's Saviours: Celebrity Conservationists in the Television Age*. London: Routledge.
John Muir Trust (2025) *John Muir Trust*. Available at: https://www.johnmuirtrust.org (Accessed: 21 June 2025).
Klapp, O. E. (1949) 'The folk hero', *The Journal of American Folklore*, 62(243), pp. 17–25.
Mackinac Center for Public Policy (2019) *The Overton Window*. Available at: https://www.mackinac.org/OvertonWindow (Accessed: 21 June 2025).
Monbiot, G. (2013) *Feral*. London: Penguin.
Paulhus, D. L. and Williams, K. M. (2002) 'The dark triad of personality: Narcissism, Machiavellianism, and psychopathy', *Journal of Research in Personality*, 36(6), pp. 556–563.
Re:wild (2025a) *Action funds*. Available at: https://www.rewild.org/action-funds (Accessed: 21 June 2025).
Re:wild (2025b) *About us*. Available at: https://www.rewild.org/about-us (Accessed: 21 June 2025).
Re:wild (2025c) *Leonardo DiCaprio*. Available at: https://www.rewild.org/team/leonardo-dicaprio (Accessed: 21 June 2025).
Roe, E. M. (1995) 'Except-Africa: Postscript to a special section on development narratives', *World Development*, 23(6), pp. 1065–1069.
Sampson, A. (2022) 'You're on the guest list: Here are the new modern status symbols', *Tatler*, 27 May. Available at: https://www.tatler.com/article/the-new-modern-status-symbols
Sandom, C. J. and Wynne-Jones, S. (2019) 'Rewilding a country: Britain as a study case' in N. Pettorelli, S. Durant, and J. du Toit (Eds) *Rewilding*. Cambridge: Cambridge University Press, pp. 222–247.
Sismondo, S. (2010) *An Introduction to Science and Technology Studies*. Chichester: Blackwell.
Soule, M. and Noss, R. (1998) 'Rewilding and biodiversity: Complementary goals for continental conservation', *Wild Earth*, 8(3), pp. 18–28.
Stoppard, T. (1993) *Arcadia*. London: Faber and Faber.
Sullivan, S. (2011) 'Conservation is sexy! What makes this so, and what does this make? An engagement with celebrity and the environment', *Conservation and Society*, 9(4), pp. 334–345.
The Ecologist (2009) *Visionaries: Derek Gow*. Available at: https://theecologist.org/2009/apr/01/visionaries-derek-gow (Accessed: 21 June 2025).
Thomas, V. (2021) *Navigating the Landscapes of Rewilding: A Comparative Case Study of 'rewilding' in the Avalon Marshes and Wild Ennerdale*. Exeter: University of Exeter.
Thomas, V. (2022) 'Actors and actions in the discourse, policy and practice of English rewilding', *Environmental Science & Policy*, 132, pp. 83–90.
Tompkins Conservation (2025) *Tompkins Conservation*. Available at: https://www.tompkinsconservation.org (Accessed: 21 June 2025).
Tree, I. (2018) *Wilding*. London: Picador.
Tree, I. (2020) 'How rewilding brought nature back to England's Knepp Estate', *Vogue*, 27 October. Available at: https://www.vogue.com/article/how-rewilding-brought-nature-back-to-englands-knepp-estate
Wade, M. (2018) 'Meet the Knepp group: The billionaires dedicated to 'rewilding' the environment', *The Times*, Available at: https://www.thetimes.com/world/us-world/article/meet-the-knepp-group-the-billionaires-dedicated-to-rewilding-the-environment-dgqrmxl9w
Weber, M. (1968) *On Charisma and Institution Building: Selected Writings*. Chicago: University of Chicago Press.

Weber, M. (1978) *Economy and Society*, G. Roth and C. Wittich (Eds) Berkeley: University of California Press. (Original work published 1922).

Weston, P. (2022) 'Knepp estate: Why the king and queen of rewilding are farming again after 20 years', *The Guardian*, 26 May. Available at: https://www.theguardian.com/environment/2022/may/26/knepp-estate-why-the-king-and-queen-of-rewilding-are-farming-again-after-20-years-aoe

Weston, R. (2020) 'It's going to be our way now': The guerrilla rewilder shaking up British farming', *The Guardian*, 4 September. Available at: https://www.theguardian.com/environment/2020/sep/04/its-going-to-be-our-way-now-the-guerrilla-rewilder-shaking-up-british-farming-aoe

Wildcard (2021) *Open letter: It's time to rewild royal land*. Available at: https://wildcard.land/campaigns/rewild-the-royals/open-letter-its-time-to-rewild-royal-land (Accessed: 21 June 2025).

Wildland (2021) *Wildland*. Available at: https://wildland.scot/ (Accessed: 21 June 2025).

Williams, E. A., Pillai, R., McCombs, K. M., and Lowe, K. B. (2021) 'The roles of character and charisma in evaluating leadership: Effects of dark triad and honesty-humility traits, attributed charisma, and COVID-19 crisis anxiety in rejecting a presidential leadership candidate', *Personality and Individual Differences*, 181, pp. 1–6.

Youens, A. (2022) 'To the eco-manor born; A new type of landowner is defying tradition to turn to nature conservation and rare breeds', *The Sunday Times*, Available at: https://www.thetimes.com/life-style/wildlife-nature/article/meet-the-landowners-rewilding-their-country-estates-8klqhkg8n

9 Rewilding and reintroductions in the Anthropocene

In discussing rewilding, this book has provided a geographical and temporal snapshot of rewilding in England and Britain in the early 21st Century. Given how rapidly rewilding is evolving, as a concept and a practice, a snapshot is not only the best that a book such as this can offer but is valuable in documenting rewilding's trajectory. In considering rewilding, this book has asked what rewilding does and can mean in Britain, where human presence and influence are persistent and pervasive. As in many other places, rewilding in Britain is contested; as rewilding negotiates and is negotiated, it is being reinterpreted in and for the British context.

Not all rewilding involves reintroductions, and not all reintroductions are rewilding, nonetheless, the two are closely linked. While the reintroduction of the red kite and the wildcat to Britain might not be considered rewilding *per se*, a consideration of them contributes to and illuminates the discourse and practice of rewilding in Britain, particularly with respect to human-animal relations. The book has illuminated these socioecological systems, together with the cultural, political, and biopolitical systems which govern rewilding. From the charismatic megafolk who influence rewilding, to the categorisation and control of the other-than-human animals involved, the power dynamics of rewilding are revealed and questioned in the hope that by acknowledging and interrogating these power structures, they might be reconsidered and reformed.

Despite the recent definition of rewilding adopted by the IUCN (2021), this book argues that rewilding cannot be viewed as a single thing that can be clearly and rigidly defined. The choice was made to characterise rewilding according to its family resemblance to a core set of traits in order to recognise its mutability and mobility – meaning different things to different people in different contexts. Family resemblance accommodates and emphasises (rather than obscuring) this diversity, which is both a strength and a weakness of rewilding; on one hand, it encourages and enables innovation and adaptation, but on the other, it can create confusion and foster controversy and concern.

Rewilding is emergent and relational, with shifting assemblages of actors, values, narratives, and interventions, contingent upon individual socioecological contexts. Family resemblance offers a useful means of understanding

DOI: 10.4324/9781003456490-10

the diverse forms of rewilding, without requiring strict definitions or static boundaries. This allows for interpretations of rewilding that are sensitive to different contexts and practices and is crucial to understanding rewilding as a dynamic and evolving practice that is shaped by social and ecological factors. This shaping is clearly evident in Britain, which has a distinct form of 'domesticated' rewilding.

This book has tried to capture some of the richness, complexity, and ambiguity of that domesticated rewilding as it is practiced, debated, and imagined in Britain today. In doing so, it has argued for an approach to rewilding that is reflexive, situated, and relational, that embraces uncertainty, engages with plural values, and foregrounds coexistence. Wildness and rewilding in Britain are involved in an ongoing negotiation and a form of rewilding is currently visible, which is navigating the tension between wild agency and human control. This book explored that tension and the way rewilding is adapting to, and reimagining, existing socioecological systems and human-nature relationships.

Domesticated rewilding can be understood on its own terms as a distinctive modality within the broader family of rewilding practices, and as a situated and pragmatic response to local constraints and opportunities. Rewilding in Britain tends to be smaller in scale than rewilding in other areas, engages with domestic as well as wild species, and permits ongoing human intervention, which in turn influences natural processes and other-than-human agency. Significantly, it also tends to self-identify as wilding rather than rewilding, reflecting discomfort with the connotations of rewilding. Rewilding is a contested and controversial concept, and domesticated rewilding is equally fraught; other scholars will argue that domesticated rewilding is inauthentic, incomplete and entails too much compromise. This book argues that the domestication of rewilding should not be seen as a weakness, rather, it may be viewed as a strength, with rewilding's plasticity allowing it to accommodate and be accommodated by other land uses. Indeed, domesticated rewilding can occur where other forms of rewilding would not be possible. A large part of this accommodation is due to, and can be seen in, human *intervention in* nature, which is partially reframed as human *interaction with* nature. Humans, and domestic animals, are involved in 'natural' processes within rewilding systems and rewilding can therefore be seen as an eco-cultural process; human and other-than-human entanglements shape, constrain and animate rewilding and reconfigure the terms of coexistence between the 'natural' and the 'cultural' worlds.

New modes of human and other-than-human coexistence is creating new modes of biopolitics. These are evident in the governance of the other-than-human animals involved in rewilding and in the governance of reintroduced species, and are emerging as part of the negotiation surrounding the restoration of wild agency. Other-than-human animals are classified, and governed, as expendable objects, machines/proxies, analogues, and self-determining agents. While visible in relation to rewilding, to a large extent, these biopolitical modes conform to conventional logics of life and death, as seen in conservation or agriculture, for example, underscoring the fact that restoring wild agency is complex and incomplete. This book highlights the negotiation of power and control and the entanglements of care and violence inherent in the biopolitics of rewilding. Certain life (of species, breeds, individuals, or even genes) is valued, nurtured and protected, while other life is made expendable, sacrificed, and even rendered killable. Decisions about which life is 'valuable' are based on shifting categories, and values which are socially, and often arbitrarily, constructed. Biopolitics provides useful insights into these power dynamics and the ethics involved in rewilding practices.

While the cats, cattle, and kites involved in rewilding and reintroductions in Britain are subject to different biopolitical modes depending on how they are classified and valued, particularly in relation to their 'wildness', it should be noted that such classifications are rarely, if ever, clear and distinct. Indeed, this book challenges the wild-domestic binary and explores different qualities of wildness. In the Avalon Marshes and the Ennerdale Valley, and in red kite and wild cat reintroductions, wildness emerges as a relational quality negotiated between humans, landscapes, and other-than-human animals.

The cattle in the Avlon Marshes and Wild Ennerdale fall into, and between, the biopolitical categories of expendable objects, machines/proxies, and analogues, occupying a liminal space between the wild and the domestic and negotiating their agency with and within biopolitical systems. Meanwhile, red kites move between the categories of charismatic wildlife on one hand and pest on the other, simultaneously switching biopolitical mode from self-determining agents to expendable objects. Cats are subject to different biopolitical modes depending on whether they are wild, feral or hybrid, with the categories of feral and hybrid blurring a possible distinction between wild and domestic. Biopolitical governance of cats is complex and highly contingent on the human actors involved, with different actors assigning different values to different cats. This is most apparent in relation to hybrids, which can be seen as problematic 'mutants', which can, and indeed should, be expended, or as agile chimeras serving as valuable analogues for the wildcat given their adaptation to anthropogenic landscapes.

Notwithstanding the agency of other-than-human animals, humans are a significant force in rewilding, with charismatic megafolk being perhaps the most significant force. The concept of charismatic megafolk offers a way of exposing and exploring the influence of these individuals in popularising rewilding, inspiring action and galvanising support. Charismatic megafolk are not simply proponents of rewilding, they control the discourse around it, forming narratives and setting the terms by which rewilding is legitimised. Even when applied benevolently, charisma's power can make it a dangerous force, obscuring complexity, concentrating authority in a small number of individuals, narrowing the discursive field, entrenching hierarchies of expertise, and reproducing power inequalities, marginalising less visible actors. When dark charisma is involved, the consequences can be even more damaging, polarising views, creating factions, and even retarding rewilding efforts.

Rewilding's charismatic megafolk, the soapboxers, trendsetters, and mavericks demonstrate clearly that rewilding is driven by emotive as well as rational decisions, and by value judgements as much as scientific evidence. This book contributes to an important discussion around the sociocultural dynamics that influence rewilding specifically but also conservation and human-nature relations more broadly. Other work is considering rewilding's stakeholders much more widely, and the importance of this is becoming increasingly evident. The communities who will be most directly affected by rewilding need to be involved in shaping rewilding practice. Understanding the needs of these communities and drawing on their diverse forms of expertise empowers rewilding to evolve according to specific ecological, social and cultural contexts, as it has in Britain and as it will continue to do.

It is important that research continues to follow rewilding's evolution, in Britain and beyond, tracing the way that its discourse and practices are shaped and adapted. In particular, the ethics of rewilding require further scrutiny; care for and responsibility to the human and other-than-human worlds affected by rewilding is under-considered in rewilding discourse and practice, including a consideration of how the costs and benefits of rewilding are borne and distributed given the prolonged time scales over which rewilding occurs. Rewilding must not only seek ecological restoration but also sociocultural restoration, and a reconsideration of human-nature relations and our environmental ethics. This is no easy task. As this book has shown, the negotiations involved have many friction points around which rewilding is moulded. These frictions are, however, generative, shaping rewilding and reintroductions, and shaping the human and other-than-human animals involved into denizens of novel, emergent socioecological systems of the Anthropocene. These socioecological systems are full of possibilities and tensions, as is already evident in the rewilding that is occurring in Britain. This book cannot, and has not, resolved these tensions. Nor does it seek to offer a manifesto for how rewilding should be. Rather, it holds up a mirror to rewilding and reintroductions as they currently are, revealing both their fallibility and their potential and offering a way of thinking about the practices, relations, and ecocultures that emerge.

Reference

IUCN CEM (2021) *Rewilding Principles*. Gland: IUCN. Available at: https://iucn.org/sites/default/files/2022-10/principles_of_rewilding_cem_rtg.pdf

Index

Pages in *italics* refer to figures, pages in **bold** refer to tables, and pages followed by "n" refer to notes.